Advanced Environmentally Friendly Technologies for Heavy Crude Oil Production

Heavy crude oil accounts for about 15–20% of total crude oil production. Despite various Enhanced Oil Recovery technologies that offer significant technical and economic success, many of these processes are not fully energy-efficient and have a considerable environmental footprint. In recent decades, several sustainable technologies and scientific advancements have emerged, offering the potential to make heavy oil and bitumen production both economically viable and more environmentally friendly. *Advanced Environmentally Friendly Technologies for Heavy Crude Oil Production* presents a comprehensive account and critical review of the technologies that have the highest potential for successful implementation in the energy transition from fossil fuels to renewable energies. It discusses progress in both academia (physical or laboratory and numerical simulations) and industry (pilots).

Features

- Emphasizes environmental aspects such as carbon intensity (kg CO_2/BBL), technology footprint, and water consumption.
- Describes the Technology Readiness Level of all routes evaluated.
- Considers techno-economic evaluation and life cycle analysis.
- Examines emerging technologies and future trends.

Professionals in the oil and gas industry as well as engineers, scientists, and regulators involved in the energy transition and carbon intensity calculations will find this book to be an excellent resource.

César Ovalles graduated with a Licentiate degree in Chemistry from Simón Bolívar University, Caracas, Venezuela, and a Ph.D. from Texas A&M University. He worked for 16 years at PDVSA-INTEVEP (R&D branch of the Venezuelan oil industry). In 2006, César joined Chevron and has been involved in R&D in petroleum chemistry, new Enhanced Oil Recovery, upgrading processes for heavy crude oils and resins, and asphaltene characterization and the mechanism of action of additives to prevent asphaltene precipitation and fouling. In 2020, he originated and led the Carbon Capture, Utilization, and Storage Program of the Chevron Technical Center in Richmond, California, until his retirement in May 2024. Dr. Ovalles has published 73 papers in peer-reviewed scientific journals and four technical books, has been awarded 25 patents, and has presented 136 papers at scientific and technical conferences.

Advanced Environmentally Friendly Technologies for Heavy Crude Oil Production

César Ovalles

CRC Press
Taylor & Francis Group
Boca Raton London New York

CRC Press is an imprint of the
Taylor & Francis Group, an **informa** business

Designed cover image: © Shutterstock

First edition published 2027
by CRC Press
2385 NW Executive Center Drive, Suite 320, Boca Raton FL 33431

and by CRC Press
4 Park Square, Milton Park, Abingdon, Oxon, OX14 4RN

CRC Press is an imprint of Taylor & Francis Group, LLC

ISBN: 978-1-041-06450-3 (hbk)
ISBN: 978-1-041-06451-0 (pbk)
ISBN: 978-1-003-63547-5 (ebk)

DOI: 10.1201/9781003635475

Typeset in Times
by SPi Technologies India Pvt Ltd (Straive)

Dedication

To Lulú, of course

To Victoria and her world to come...

Contents

Foreword

A recognized source on fossil fuels global production scenarios suggests the existence of a peak occurring around 2030, spanning 106 ± 3 million barrels per day (MMB/day); then going down to 86 ± 14 MMB/day by 2050 (McKinsey & Co 2025). Under the most optimistic outlooks considered in this report, oil will represent 41–55% of the world's energy sources by 2050, with oil being the most important energy source among fossil fuels (i.e., coal, oil, and gas) in all scenarios. The preceding projections are similar to those disclosed by other essential sources of information on global energy (IEA 2025 and DNV 2025), indicating that fossil fuels, petroleum among them, will still be vital to achieve the world's energy demands in 2050. By pooling heavy oils (HO) and extra-heavy oils (XHO) (or "bitumens") as the joint category "HO" which shows specific gravities <20°API, many countries in the world have proven reserves of HO, particularly Canada and Venezuela. Since about 20% of the world's daily production of petroleum has been HO, these materials will continue to be significant energy sources for the upcoming decades.

Since HOs production is more labor-, investment-, energy-, and pollution-intensive compared to conventional petroleum production, improved alternatives to those developed during the past decades are required for coping with economic and environmental requisites, which keep continuously becoming more challenging. It is the purpose of the present monograph to cover in greater detail such advanced, economical, and environmentally friendly technologies for HO production. A brief description of the technologies, their achievements, and future expectations is addressed, with important references appended for perusal of interested readers. Very important to consider is the fact that comparison of the discussed technologies is made on the basis of their Technology Readiness Level (TRL) scale vs. their potential greenhouse gas (GHG) emission reduction. The preceding provides a broad and meaningful tool for identifying the most important qualifiers of the covered technologies. A thorough set of technologies is covered in the monograph, divided into non-thermal, thermal with/without inclusion of catalysts, and a series of novel emerging technologies, among which nanomaterials, ionic liquids, biomaterials, ultrasound, radiofrequencies, and subsurface hydrogen production are included.

The present monograph is a very well-conceived review of the state of the art regarding recent environmentally friendly enhanced recovery-upgrading technologies for HO. It is deemed a good source of actualized information for scientists, engineers, and practitioners of enhanced HO recovery under more feasible, economical, and non-pollutant technologies. Personally, this monograph is considered of utmost importance for students and newcomers to the field of enhanced production-HO upgrading, with particular focus on new-emerging subsurface alternatives.

From the preceding, I thank Dr. Ovalles for compiling such an important document and lend my personal support to the widespread distribution of this rich source of information.

Sincerely,

Lante A. Carbognani Ortega
Researcher, Laboratory Manager
NANOS TECH Ltd. Calgary, AB-Canada

REFERENCED DOCUMENTS

DNV 2025. Energy Transition Outlook 2025. https://www.dnv.com/energy-transition-outlook/download/ (Accessed on December 12th, 2025)

IEA 2025. Global Energy Review 2025. https://www.iea.org/reports/global-energy-review-2025 (Accessed on December 12th, 2025)

McKinsey&Co 2025. Global Energy Perspective 2025. https://www.mckinsey.com/industries/energy-and-materials/our-insights/global-energy-perspective#/ (Accessed on December 12th, 2025)

Preface

Currently, the world's reserves of heavy crude oil and bitumen (HCO/B) make up over 50% of all hydrocarbons globally and account for about 15–20% of total crude oil production. Since the last century, various Enhanced Oil Recovery (EOR) technologies have been utilized with significant technical and economic success. However, many of these processes are not fully energy-efficient, produce a substantial amount of contaminants and greenhouse gases (GHGs), require high water consumption, and have a considerable environmental footprint. In recent decades, several sustainable technologies and scientific advancements have emerged, offering the potential to make heavy oil and bitumen production both economically viable and environmentally friendly.

The objectives of this monograph are to provide a comprehensive account and critical review of the environmentally friendly technologies for heavy crude oil (<20°American Petroleum Institute [API]) recovery that show the greatest potential for success in the energy transition from fossil fuels to renewable energies. We will discuss advancements in academia, including physical experiments and numerical simulations, as well as industry developments, particularly pilot projects and efforts related to environmental sustainability. This information will be drawn from patents and open literature up to the date of submission of this book to the publisher. Our aim is for engineers, scientists, and practitioners in the field to utilize this book as a valuable resource and reference in their research and technology development. Additionally, chemical, petroleum, and mechanical engineering students can use this book for their coursework and research related to their graduate theses.

Chapter 1 provides a comprehensive assessment of technology based on its deployment potential. It evaluates factors such as technical scalability, reduced associated risks, and the capability to lower GHG emissions compared to the base cases—waterflooding for non-thermal processes and steamflooding for thermal technologies.

Chapter 2 examines non-thermal EOR technologies for extracting heavy crude oils and bitumens. These production methods have been successfully implemented with a Technology Readiness Level (TRL) of 9 due to their relative simplicity and cost-effectiveness. Typical recovery factors range from 5% to 12% of the Original Oil in Place (OOIP). However, several technological approaches are emerging that can significantly impact GHG emissions and make conventional non-thermal HCO/B production processes more environmentally friendly, while maintaining or improving oil recovery factors up to 30% of the OOIP. This chapter describes recent developments in cold production (multilateral wells, electrical downhole heaters, and enhanced cold heavy oil production with sand), immiscible gas injection (carbon dioxide and nitrogen injection processes), chemical EOR (polymer flooding, alkaline-surfactant-polymer flooding, and emulsion injection), solvent-only injection (hydrocarbon and environmentally friendly solvents), and subsurface solvent deasphalting.

Chapter 3 focuses on thermal methods for producing HCO/B. These technologies utilize heat to reduce the viscosity of oil, facilitating extraction from reservoirs where non-thermal methods have been exhausted, proven ineffective, or are economically unfeasible. The most widely used thermal methods are hot-water injection, cyclic steam injection, steamflooding, and Steam-Assisted Gravity Drainage. However, these methods face significant challenges, such as high energy and water consumption, the need for costly water treatment to generate steam, substantial heat losses, and considerable GHG emissions. Over the last two decades, several innovative technological advancements have significantly reduced GHG release, making conventional thermal heavy oil production methods more environmentally friendly while achieving recovery factors of **30–70%** of the OOIP. This chapter explores recent developments in low-carbon-intensity (CI) steam injection technologies, the co-injection of non-condensable gases, electromagnetic heating, and the utilization of low-CI energy sources for HCO/B production, including solar steam generation, nuclear-powered EOR, and geothermal energy.

Chapter 4 discusses several emerging technologies for EOR of HCO/B production that are coming over the horizon, with the potential to increase oil rates and recovery factors while simultaneously reducing GHG emissions, technology footprint, and water consumption. The technologies evaluated include nanomaterials, ionic liquids, microbial- and plant-based biomaterials, ultrasound applications, as well as subsurface hydrogen production, and switchable surfactants.

Chapter 5 presents carbon intensities and GHG emission reductions associated with various conventional and emerging technologies for EOR of HCO/B. The data indicate that the non-thermal technologies with the highest percentage of GHG emission reduction are Cyclic Solvent Injection using Biodegradable Solvents, Multilateral Wells and Autonomous Inflow Control Devices (AICD), Nanomaterials, and Polymer Flooding. The thermal routes with the highest GHG abatement are radiofrequency (RF) heating, Enhanced Modified Steam and Gas Push (eMVAPEX), Enhanced Solvent Extraction Incorporating Electromagnetic Heating ("ESEIEH"), Nuclear Energy EOR/Steam, Geothermal/Steam, Solar Energy EOR/Steam, Solvent-SAGD Co-Injection, and Ultrasound Energy.

It is important to emphasize that the results presented are a high-level comparison of the most environmentally friendly technologies. Some processes are well researched and capable of producing large volumes of oil, while others are based on limited data and apply to only a small portion of the resource base. It is essential to analyze each project individually, considering factors such as location, reservoir characteristics, availability of raw materials, facility design, production duration, and the company's expertise, among other relevant aspects. Additionally, Chapter 5 outlines future technological trends, including hybrid EOR concepts and potential research and development directions for the next decade.

It can be foreseen that petroleum will continue to play a significant role in our lives in the near- to medium-term future. The production of HCO/B will remain strategically and commercially important to meet the world's energy demands in 2050 and beyond. Even if the world transitions to energy sources independent of HCO/B, various petroleum-derived feedstocks remain essential for the production of everyday products such as plastics, pharmaceuticals, and cosmetics. This dependence on petroleum for manufacturing is yet another reason not to use it solely as a fuel.

From a chemical characterization standpoint, HCO/B has a relatively high carbon-to-hydrogen molar ratio. When burned, it produce more carbon dioxide than conventional crude oils and natural gas. Therefore, heavy crude oil and bitumen are likely better suited as feedstocks for petrochemicals or asphalt products rather than for fuel.

Acknowledgments

Many thanks to Ian Benson and Lante Carbognani for their feedback and recommendations throughout the development of this monograph. Their support was indispensable in delivering work of truly world-class quality.

The author gratefully acknowledges the insightful technical discussions and debates with colleagues and friends Estrella Rogel, Leonor Rouse, Berna Hascakir, Francisco Lopez-Linares, Babak Fayyaz-Najafi, Juan D. Muñoz, Pedro Vaca, Jeff Woolford, Gunther Dieckmann, Don Kuehne, Michal Okoniewski, and Art Inouye. To all, the author's most profound gratitude.

My gratitude to Gisela Cardozo, Reina Valbuena, and Humberto Sanchez for their help in designing the book's cover.

Finally, the author wishes to thank all the individuals he has worked with, shared experiences with, and interacted with throughout his over 40-year career in the petroleum industry in Venezuela and the USA. He has enjoyed every moment of it and hopes this journey never ends.

1 Introduction

The world's reserves of heavy crude oil and bitumen (HCO/B) constitute more than 50% of all hydrocarbons on the planet and account for approximately 15–20% of total crude oil production. Global production of HCO/B is projected to increase from 2025 into the near future (e.g., to 2030), primarily driven by Canada, the USA, and Brazil (IEA 2025). These feedstocks have very high viscosity, a low amount of distillable materials (such as gasoline or diesel), and high levels of contaminants, including sulfur, nitrogen, and heavy metals like vanadium and nickel. As a result, their production, transportation, and refining present significant challenges and costs using conventional technologies.

In April 2016, nearly every country in the world signed the Paris Agreement, which aims to keep global warming well below 2°C above pre-industrial levels and to pursue efforts to limit it to 1.5°C (United Nations 2016). Since that time, only six countries have reduced their emissions in line with their pledges, while the USA has once again exited the Agreement. It is now widely recognized that the world is unlikely to achieve net-zero emissions by 2050 (DNV 2025 and McKinsey 2025). Additionally, DNV's annual Energy Transition Outlook projects a shift from the current energy mix of 80% fossil fuels and 20% non-fossil fuels to a balanced 50/50 mix by 2050 (DNV 2025). Therefore, the production of HCO/B will remain strategically and commercially important to meet the world's energy demands in 2050 and beyond.

Since the 20th century, various Enhanced Oil Recovery (EOR) technologies have been employed with technical and economic success. However, many of these processes are not energy efficient, produce a substantial amount of contaminants and greenhouse gases (GHGs), require high water consumption, and have a significant environmental footprint. In the last decade, several sustainable technologies and scientific advancements have emerged, offering the potential to make heavy oil production not only economically viable but also environmentally friendly. Before entering the technology assessment, this chapter will define and present the world's reserves of HCOs/B, discuss current issues with conventional production technologies, outline objectives, and clarify what are in-scope and out-of-scope.

1.1 DEFINITIONS AND WORLD RESERVES OF HEAVY CRUDE OILS

According to the International Energy Agency (IEA), fossil fuels met 80% of global energy demand in 2023. IEA's scenarios showed that demand for crude oil, natural gas, and coal will peak by 2030, although oils used for aviation and petrochemicals will increase by 2050 (International 2024 and IEA 2025). In general terms (Ovalles 2019), crude oils are defined as light (more than 31.1°American Petroleum Institute [API]), medium (between 20.1° and 31°API), heavy (between 10° and 20°API), and extra-heavy or bitumen (less than 10°API), where API is defined according to equation 1.1.

DOI: 10.1201/9781003635475-1

$$\textbf{API Gravity} = \frac{\textbf{141.5}}{\textbf{Sp.Gr.60F/60F}} - \textbf{131.5} \tag{1.1}$$

where Sp. Gr. 60F/60F is the specific gravity of the crude at 60°F vs. water at the same temperature.

The 2nd International Conference on Heavy Crude and Tar Sands (UNITAR 1982) proposed a more detailed definition for heavy- or extra-HCOs and bitumen based on API gravity and viscosity (**Figure 1.1**). As before, HCOs have an API between 10°API and 20°API but have a gas-free viscosity at the original reservoir temperature between 100 and 10,000 mPa s (cP). Additionally (**Figure 1.1**), extra-heavy and bitumen have an API gravity lower than 10°API and a gas-free viscosity at the original reservoir temperature greater than 10,000 cP. These definitions have received worldwide acceptance since 1982. For simplicity, **in this book, the term "heavy crude oil" is used in the broad sense of petroleum-containing oils with an API gravity lower than 20°API and a reservoir viscosity higher than 100 mPa·s**.

HCOs are some of the most abundant energy resources on the planet. The world reserves of HCO represent more than 50% of all hydrocarbons available on earth, with more than 8,900 billion barrels (BBBL) of original oil in place (OOIP) (British Petroleum 2016 and Guzman 2015). **Figure 1.2** shows the geographic distribution of HCOs worldwide. South (38%) and North (34%) Americas have the most extensive heavy oil volumes, totaling 6,429 BBBL of OOIP. The Middle East (11%), Russia (6%), and Transcaucasia (5%) have over 1,000 BBBL of OOIP, followed by Africa (2%), Europe (1%), and South East Asia and Oceania (2%) (Meyer 2007; Guzman 2015).

More than 30 countries have heavy or extra-heavy crude oil and bitumen. Venezuela has proven reserves of ~1,000 BBBL, mainly located at the Orinoco Oil Belt (north of the Orinoco River), the Bolivar Coastal Field, and the Boscan Field (west of the Maracaibo Lake) (Schenk 2009). Canada has estimated HCO reserves of ~1,000 BBBL, of which more than 20% are surface mineable bitumen, and the remaining 80% are considered "*In-situ* bitumen reserves" (Gray 2015). The latter are too deep to mine and can only be produced using thermal processes such as

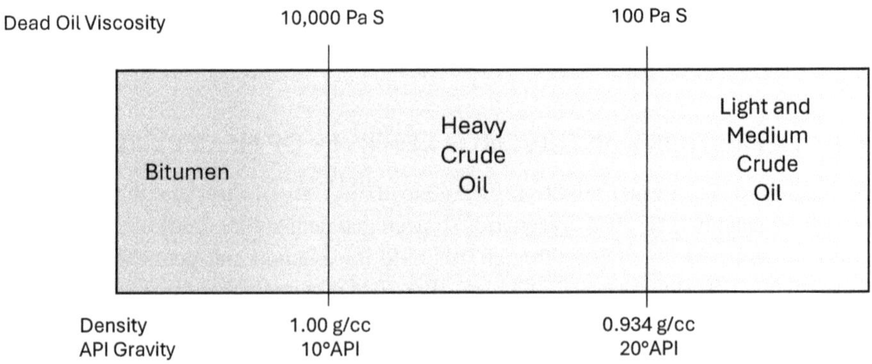

FIGURE 1.1 Definition of natural bitumen, heavy, medium, and light oil based on viscosity, density, and American Petroleum Institute (API) gravity (Ovalles 2019).

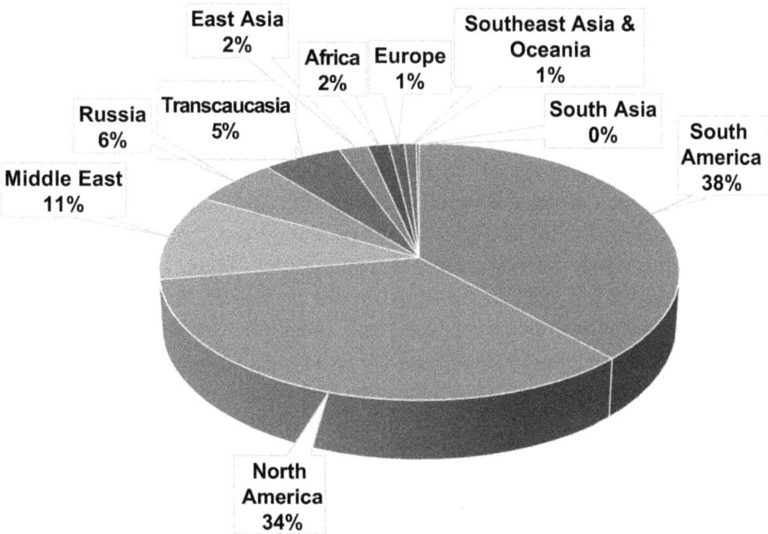

FIGURE 1.2 Geographic distribution of heavy crude oils (HCOs) (American Petroleum Institute [API] lower than 20°API) worldwide in billions of barrels of oil of original oil in place (OOIP).

Data taken from (Guzman 2015).

Steam-Assisted Gravity Drainage (SAGD), among other technologies (Gray 2015). Other significant Canadian bitumen accumulations include Cold Lake and Peace River deposits.

California, USA has considerable reserves of heavy oil located at Midway-Sunset (over three BBBL), Wilmington (approximately three BBBL), Kern River (over three BBBL), and South Belridge (slightly less than two BBBL) fields (Atwater 2020). Three of the world's largest HCO oilfields (>1,000 million BBL of OOIP) are situated in Al-Burqān in Kuwait, Kirkuk in Iraq, and AbūSaʿfah in Saudi Arabia. Other giant fields producing heavy oil include Duri in Indonesia, Gudao and Karamai in China, Ebano-Panuco in Mexico, Belayim Landin in Egypt, Maydan Mahzam in Qatar, and Uzen and Zhetybay in Kazakhstan. Additional bitumen is located in Russia, mainly in the Volga-Urals and Siberian regions, and in Kazakhstan (Atwater 2020).

The proven and estimated reserves of HCOs (API lower than 20°API) are vast and can be found on all continents except Antarctica. Thus, it can be safely forecasted that these raw materials will be key components of the energy mixture throughout the 21st and 22nd centuries (Ovalles 2019). Some challenges of current technologies in the exploitation of HCO are described next.

1.2 ISSUES WITH CURRENT PRODUCTION TECHNOLOGIES

In the last 45 years, several EOR technologies have been utilized with technical and economic success for the production of (Huc 2011, Ovalles 2019, Islam 2020, and Ahmadi 2023). Unfortunately, many HCOs of these processes are not energy efficient, generate a relatively large amount of contaminants (SO_x and NO_x) and GHGs, have high water consumption, and have a relatively large footprint.

As known in the literature, HCOs are reasonably well chemically characterized, but complex materials (Altgelt 1994, Strausz 2003, Speight 2007, Boduszynski 2015, Ramirez-Corredores 2017, and Ovalles 2019, Chapter 2). They have a very high viscosity, a low amount of distillable fractions (gasoline or diesel), and high levels of undesirable constituents such as sulfur, nitrogen, and heavy metals (mostly vanadium, nickel, and iron) that make their production, transportation, and refining very difficult and costly by conventional technologies. Thus, HCO properties must be improved before the oil can be transformed into fuels, lubricant oils, plastics, petrochemicals, etc. Several partial, medium, and high conversion upgrading technologies have been developed to address those issues at the refinery level. The main challenge is to move the viscosity reduction and contamination removal upstream and midstream with the concomitant saving in capital expenses (CAPEX) and operating expenses (OPEX) (Ovalles 2019).

In general, HCO production has a more significant impact on the environment compared with those for medium and light crude oils. The energy consumed to lift and handle a barrel of heavy oil is between 20% (Smalley 2000) and 33% of the oil produced (Gordon 2015). This energy consumption is several times greater than that needed to produce a barrel of medium or light oil. Also, in conventional HCO technologies, 85% of GHG emissions are due to the combustion of natural gas for steam and power generation (Huc 2011).

Figure 1.3 illustrates the comparison of upstream GHG emissions in kilogram of carbon dioxide equivalent removed for one US Barrel of petroleum (kg CO_2e/BBL) for selected light (L.) crude oils, medium (M.) crude oils, and HCOs. The data show that the average carbon intensity (CI) for HCO is 80 kg CO_2e/BBL, which is significantly higher than the emissions from light and medium crude oils, which average 40 kg CO_2e/BBL. Additionally, higher values are reported for average HCOs from California, Canada, Venezuela, and Indonesia, ranging from 148 to 190 kg CO_2e/BBL (Gordon 2015, El-Houjeiri 2017, Beckman 2018, Masnadi 2018, Canbaz 2021, Lee 2021, and Carbon 2025).

Furthermore, Gordon et al. modeled 30 global test oils and found that the heavier the oil, the more heat, steam, and hydrogen are required to extract, transport, and transform it into high-value petroleum products like gasoline and diesel (Gordon 2015). However, different classes of oils generally result in varying levels of total GHG emissions. For instance, California's HCOs necessitate the injection of large volumes of steam into the reservoir before any petroleum products can reach consumers. Generating this steam consumes up to one-third of the energy content of the oil, primarily in the form of natural gas. Similar scenarios are observed worldwide, particularly with heavy oils that have high water cuts and high gas production (Gordon 2015).

Moreover, conventional steam injection techniques suffer from poor vertical sweeping related to steam gravity in thick-layer reservoirs, severe heat loss to overburden in thin-layer reservoirs, and steam channeling between the injector and producer (Gomaa 2024). Technological advancements aimed at improving energy efficiency hold the greatest potential for narrowing the GHG intensity gap between

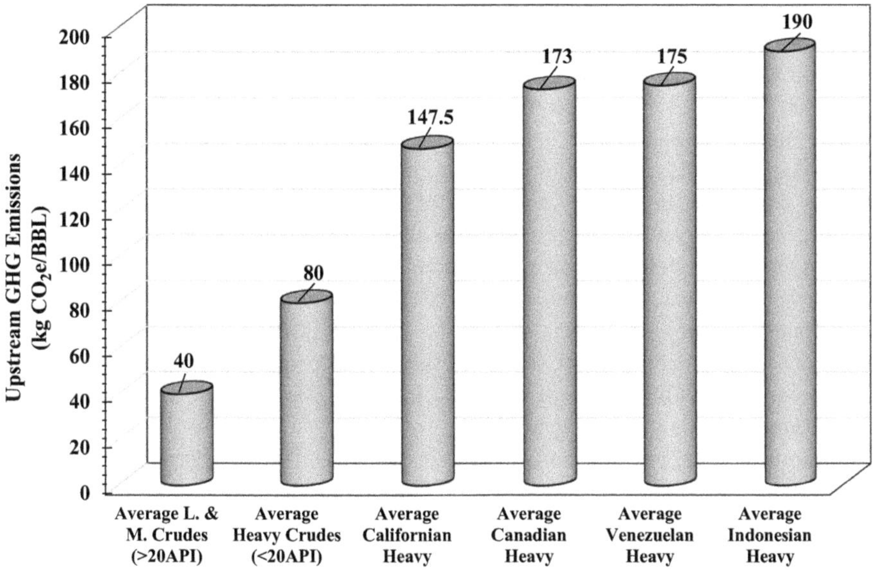

FIGURE 1.3 Comparison of upstream greenhouse gas (GHG) emissions in kilogram of carbon dioxide equivalent removed for one US Barrel of petroleum (kg CO_2e/BBL) crude for selected light (L.) crude oils, medium (M.) crude oils, and heavy crude oils (HCOs).

Data taken from (Gordon 2015, El-Houjeiri 2017, Beckman 2018, Masnadi 2018, Canbaz 2021, Lee 2021, and Carbon 2025).

HCO and conventional production methods (Bohm 2012). The current regulatory environment can potentially make heavy oil portfolios less competitive than those based on conventional and lighter crudes (Lee 2021). Moreover, in upcoming regulatory scenarios, HCO producers might need to lower their crude prices to compete with lower carbon-intensity alternatives, which increases the risk of becoming economically unviable.

Another important environmental aspect of heavy oil production is the relatively higher water consumption and handling compared to the production of medium and light crude oils (Rosa 2017). The need to use steam-injection processes for enhanced heavy oil recovery requires a constant source of freshwater since salt-containing aqueous fluids cannot be used because of severe, irreversible damage to the boilers (Donaldson 1979 and Morrow 2014). For instance, in the Kern River oilfield in California, two barrels of water are required for each barrel of oil produced. Also, in Alberta, up 80% of the water can be recycled while the rest has to be fresh water (Canbaz 2021).

In all cases, the demand for freshwater must be compatible with regional demand from municipalities and agricultural interests. On the other end of the processes, large amounts of produced water must be separated from the oil, treated, and disposed of in accordance with existing pollution-control regulations (Donaldson 1979 and Rosa 2017). In turn, this situation increases the cost and environmental concerns of conventional heavy oil technologies.

Furthermore, heavy oil production facilities have a relatively sizeable area footprint. Because oil flow rates are generally low, many more wells are required to develop an HCO field than the equivalent light and medium oil counterparts (Smalley 2000). In relatively unexplored and environmentally sensitive areas, such as the Arctic tundra, South America's rain forest, or wetlands, the use of horizontal well technology has reduced the number of wells, allowing several wells to be drilled from a single pad. In this way, the effect of HCO production on the environment is minimized. However, this topic should be addressed adequately to avoid permanent damage to the ecosystem using conventional HCO production technologies.

Furthermore, heavy oil extraction and other forms of industrial development have contributed to significant forest loss. From 2000 to 2012, Alberta lost 5.5% of its total land area to deforestation. Similar levels of deforestation have been reported in other countries, with Brazil experiencing a loss of 4.3%, the USA 2.9%, and Russia 2.2% (Petersen 2014 and Canbaz 2021).

Despite all the previous challenges, in the last 10 years, several sustainable technologies and scientific developments have come over the horizon that make HCO production a techno-economic success and environmentally friendly. This area is the focus of this book.

1.3 OBJECTIVES

Our objectives are to provide an "in-depth" account and critical review of the environmentally friendly technologies for HCO (<20°API) recovery that show the greatest potential for success in the energy transition from fossil fuels to renewable energies. We will discuss the progress of academia (physical or laboratory and numerical simulation), industry (pilots), and environmental/sustainability efforts as reported in the patent and open literature up to the moment of submitting this book to the publisher. The idea is that engineers, scientists, and practitioners in the area use this book as a resource and reference during their research and technology development work. Chemical, petroleum, and mechanical engineering students can also utilize this book in their coursework and research for their graduate theses.

Our goal is to compare HCO recovery methods based on a combination of CI (e.g., \$/Ton CO_2 abated or kg CO_2e/BBL), water usage, and other metrics such as aerial footprint, and environmental concerns. We aim to identify the most impactful and promising approaches. In the following section, we will outline the criteria and methodologies used for selecting these technologies, along with a comparison to conventional methods.

Our long-term objective is to show that producing heavy or extra-heavy crude oils and bitumen in an environmentally conscious way is vital for the oil and gas industry. In our opinion, these essential raw materials represent the transition from a fossil fuel-based economy to a renewable-energy-based society.

1.4 WHAT ARE IN-SCOPE AND OUT-OF-SCOPE?

Table 1.1 presents the list of in-scope and out-of-scope topics for environmentally friendly EOR technologies for heavy oil production. As mentioned in the previous

TABLE 1.1
In- and Out-of-Bound Topics for Environmentally Friendly Enhanced Oil Recovery (EOR) Technologies for Heavy Oil Production

In-Bound Topic	Out-of-Bound Topic
• Heavy oil reservoirs ($<20^{\circ}$API). • Non-thermal production methods, that is, recent development in cold production, solution–gas drive mechanism, immiscible gas injection (N_2 and CO_2), conventional and biodegradable solvent-only injection, chemical EOR, and subsurface deasphalting. • Thermal and hybrid EOR methods such as low CI steam injection, solvent, chemical additives co-injection, as well as Non-Condensable Gases (NCGs) and steam/flue gas co-injection. • Emerging EOR technologies, that is, Nanotechnology and nanomaterials, use of ionic liquids, microbial- and plant-based sustainable materials, and subsurface hydrogen production. • Different sources of energy (steam, solar, geothermal, radiofrequency, nuclear, and ultrasound). • Environmental considerations, that is, advantages and disadvantages (leaving contaminant behind, CO_2 sequestration, etc.)	• Medium, light, or condensate reservoirs ($>20^{\circ}$API). • Tight or shale oils. • Chemical and molecular characterization of HCOs and bitumen. • First-principles fundamentals of heavy oil recovery and fluid phase behavior. • Chemistry and mechanistic pathways for HCO/B conversion. • Conventional *In-situ* Combustion. • Any explosion-related downhole energy source using conventional or nuclear explosives.

sections, this book focuses on heavy oils, which are defined as API gravity lower than 20.

It is considered **in-scope** the non-thermal and thermal production methods for Improved Oil Recovery (IOR), aimed at increasing the amount of HCO that can be extracted from the reservoir. For example, recent developments in cold production include solution–gas drive mechanism, immiscible gas injection (N_2 and CO_2), conventional and biodegradable solvent-only injection, chemical EOR, and subsurface deasphalting. Additionally (**Table 1.1**), it will be considered thermal and hybrid EOR methods such as low CI steam injection, solvent, chemical additives co-injection, as well as NCGs and steam/flue gas co-injection, as well as emerging EOR technologies such as nanotechnology and nanomaterials, use of ionic liquids, microbial- and plant-based sustainable materials, and subsurface hydrogen production. Finally, different sources of energy (steam, solar, geothermal, radiofrequency, nuclear, and ultrasound) and environmental considerations will be examined.

Furthermore, several environmentally friendly technologies are coming over the horizon and are considered in-scope (**Table 1.1**). Examples are nanotechnology, ionic liquids, renewable materials, subsurface hydrogen production, *In-situ* Combustion with Carbon Capture and Storage, and ultrasound applications. Different energy sources, such as "green" energy, electrical, and radiofrequency, will also be considered.

Special focus will be made on environmental aspects such as CI (as measured in kg CO_2e/BBL or kg CO_2e/MJ), technology footprint, water consumption, use of specialty chemicals and additives, and safety considerations (potential for spills, loss of containment, etc.). Techno-economic evaluation and life cycle analysis will be included when the data are available. Also, surface facilities and well designs that affect the CI of the process will be discussed.

The literature review will include general descriptions of each technology's fundamentals, physical or laboratory experiments, numerical simulations, and field tests in the area up to the moment of submitting this book to the publisher. The advantages and disadvantages of each route will be presented so the reader can judge the merits and potential for such routes.

Conversely (**Table 1.1**), it is considered out-of-scope chemical characterization of HCOs/B, as well as detailed first-principles fundamentals of heavy oil recovery and fluid phase behavior. These topics have been extensively reviewed in the literature (Altgelt 1994, Speight 2007, Ovalles 2019, Lee 2021, Islam 2020, and Ahmadi 2023). Other topics that will not be examined are conventional *In-situ* Combustion and the use of any explosion-related downhole energy source, such as conventional or nuclear explosives. Finally, proprietary information, ideas, and technologies not openly available are out-of-scope and will not be presented.

1.5 HIGH-LEVEL TECHNOLOGY ASSESSMENT

In broad terms, technology comparisons and evaluations can be a complex endeavor as it depends on data availability and are subject to change as new information continuously emerges. Also, different stakeholders may have other priorities and perspectives, leading to subjective evaluations (Taber 1997 and Vishnumolakala 2020). Therefore, it is essential to provide a high-level selection of the most promising technologies to highlight viable technological pathways, emphasizing their fundamentals and scalability from laboratory settings to pilot projects and full deployment. Detailed descriptions of all of the heavy or extra-HCOs and bitumen production technologies are described elsewhere (Huc 2011, Gray 2015, Ovalles 2019, Strausz 2003, Islam 2020, and Ahmadi 2023).

The Technology Readiness Levels (TRLs) are a valuable tool for identifying the stage of development of the technology associated with a particular project, as well as the planned progression during the project execution (Department 2015). In the downstream sector (which is not the focus of our discussion), a technology typically progresses from laboratory development to bench testing, then to pilot testing, demonstration, and finally to commercial deployment (Euhus 2021).

In contrast, the upstream sector (which is our focus) begins with laboratory testing, also known as "physical simulation," followed by modeling or numerical simulation (representing TRL levels between 1 and 4). Next, the technology undergoes field piloting under realistic conditions (TRL levels between 5 and 6), followed by a demonstration phase (TRL levels between 7 and 8) that includes limited or early adoption. Often, there is a preference to transition directly from the field pilot stage—typically involving one or two wells—to the deployment phase, which encompasses multi-well and multi-field development (TRL level 9). A summary of

TABLE 1.2
Upstream Technology Readiness Levels (TRLs) Used in This Book

TRL Number	Description
1	Basic principles observed or demonstrated in the literature.
2	Proof of concept or proof of principle demonstrated in a laboratory or literature.
3	Proven in laboratory scale, also known as physical simulation.
4	Successful modeling or numerical simulation. Matched experiments.
5	Initial field test under favorable or near-realistic conditions. Most of the components were successfully tested.
6	A successful field test was conducted under realistic conditions. All components integrated and successfully tested.
7	Demonstration, limited adoption either in several wells or in multi-field development. Small-scale field pilot.
8	Demonstration, early adoption in multi-well or multi-field development. Commercial-scale field pilot.
9	Successful deployment in multi-well and multi-field developments worldwide.

Source: Scale based on the Department of Energy (Department 2015) with few modifications.

the upstream TRLs used in this book is presented in **Table 1.2**. This scale was based on the Department of Energy descriptions (Department 2015) and used for all the cases examined with few modifications.

Figures 1.4 and **1.5** show the potential for GHG emission reduction vs. the TRL. These illustrations present a high-level comparison of non-thermal and thermal HCO production technologies. The y-axis represents the potential GHG emission reductions of various technologies compared to a Base Case. Specifically, "Low" indicates a decrease of less than 30%, "Medium" signifies a reduction between 30% and 70%, and "High" reflects reductions greater than 70%, vs. the Base Case (see below). This is a qualitative ranking, and it is based on available literature. On the other hand, in terms of CI, it has been reported that a "Low" ranking corresponds to potential emissions of lower than 30 kg CO_2e per barrel (BBL) of crude oil, "Medium" indicates emissions in the range of 30–60 kg CO_2e/BBL, while "High" is attributed to values exceeding 60 kg CO_2e/BBL (Carbon 2025). Broadly, the lower the CI of a technology, the higher its potential to reduce GHG emissions. Both rankings are used in this work.

For non-thermal technologies (**Figure 1.4**), the Base Case is waterflooding; for thermal processes (**Figure 1.5**), the Base Case is steamflooding. As anticipated, the extent of GHG reduction varies based on factors such as field characteristics, field maturity, depth, thickness, petrographic properties, geographical location, facility design, and production duration. It is important to note that other GHG emissions, such as methane releases, are not considered in this analysis, which focuses solely on subsurface technologies.

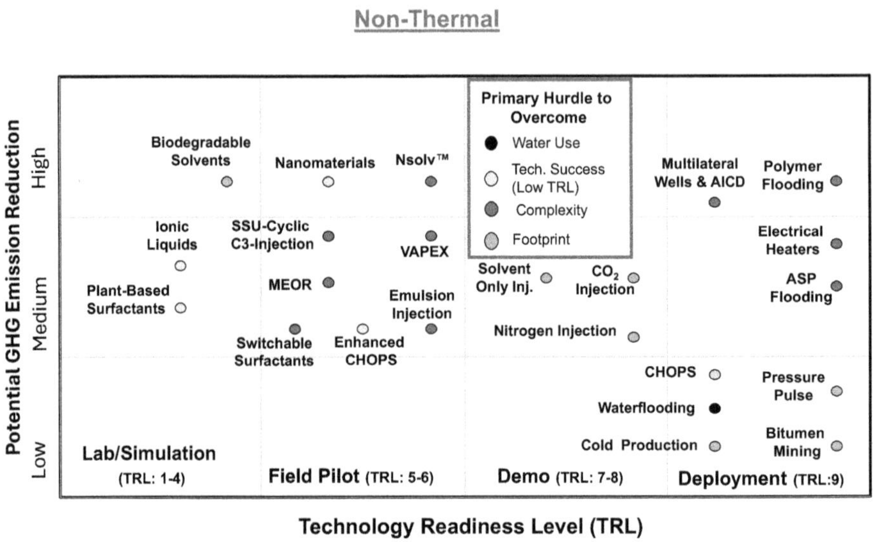

FIGURE 1.4 Potential for greenhouse gas (GHG) emission reduction vs. Technology Readiness Level (TRL) for **non-thermal** heavy crude oil (HCO) production technologies. Abbreviations: AICD: Autonomous Inflow Control Devices; ASP: Alkaline-Surfactant-Polymer; CHOPS: Cold Heavy Oil Production with Sand; MEOR: Microbial-Enhanced Oil Recovery or Microbial-Derived Surfactants and Polymers; Nsolv™: Warm Solvent Injection; SSU: Subsurface Upgrading via Cyclic Propane Injection; and VAPEX: Vapor Extraction Process.

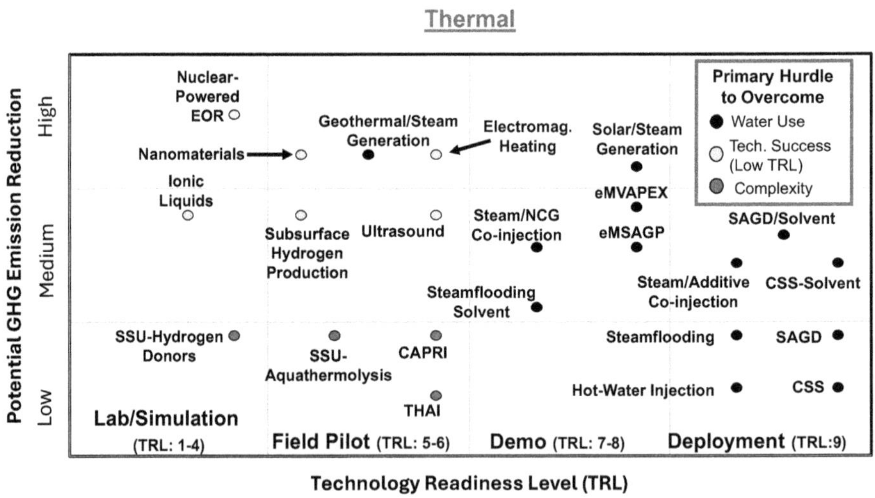

FIGURE 1.5 Potential for greenhouse gas (GHG) emission reduction vs. Technology Readiness Level (TRL) for **thermal** heavy crude oil (HCO) production technologies. Abbreviations: CAPRI: CAtalytic upgrading Process; CCS: Carbon Capture and Storage; CSS: Cyclic Steam Stimulation; eMSAGP: Enhanced Modified Steam and Gas Push; eMVAPEX: Enhanced Modified Vapor Extraction Process; NCG: Non-Condensable Gas (natural gas or flue gas); SAGD: Steam Assisted Gravity Drainage; SSU: Subsurface Upgrading; THAI: Toe-to-Heel Air Injection.

Additional information has been incorporated into **Figures 1.4** and **1.5**, located in the top right corner. This information highlights the primary challenge that a technology must address in order to succeed in the current energy transition scenario. In the insert, water consumption is represented by black circles, indicating the main challenge for methods that rely on the downhole injection of this liquid. Technical success is depicted with yellow circles, which signify the relatively low TRL of the technology and the necessity to reduce risks as it progresses to higher TRL values. Complexity is illustrated with red points, showing that the use of specialty chemicals and additives and multiple steps are required both at the subsurface level and/or in surface installations. Lastly, the high footprint is represented by orange dots, indicating that the space occupied by the process's surface facilities is a significant concern.

1.5.1 Non-Thermal Technologies

Non-thermal technologies for producing HCOs do not rely on the addition of heat to the mineral formation for extraction but instead utilize the internal energy of the reservoir (Mai 2009). These methods are commonly used for extracting HCOs due to their simplicity and cost-effectiveness. Among these technologies are conventional cold production (without sand), Cold Heavy Oil Production with Sand (CHOPS), waterflooding, bitumen mining, and pressure pulse. As illustrated on the right side of **Figure 1.4**, these technologies have been successfully implemented (TRL level 9, see **Table 1.2**) for HCO production with multi-well and multi-field applications in different parts of the world during the 20th century. Recovery factors typically range from 5 to 12% of the OOIP. However, due to their maturity and high degree of optimization, continuous improvement on these routes has a relatively low potential for reducing GHG emissions in the future (Lee 2021). One of the main challenges they face is a relatively larger environmental footprint compared to facilities for medium and light crude oils (Smalley 2000, Canbaz 2021, and Morales 2024). In the case of waterflooding, water consumption is a significant concern (Rosa 2017). However, there are several emerging technological approaches that can "move the needle" from "Low/Medium" to "Medium/High" GHG emission reduction and make those conventional non-thermal heavy oil production methods more environmentally friendly, maintaining or improving the oil rates and recovery factors. These routes are discussed in this monograph, and they are Multilateral Wells and Autonomous Inflow Control Devices (AICD) in Section 2.1.1), Electrical Downhole Heaters (Section 2.1.2), Enhanced CHOPS in Section 2.1.3, and Solvent-Only Injection (Section 2.4).

Furthermore, based on the criteria outlined above, we selected the other technological routes that show medium to high potential for reducing GHG emissions vs. the Base Case (in this case, cold production), as illustrated in **Figure 1.4**. We identified environmentally friendly pathways that are more likely to continue or be deployed due to their technical scalability, economic potential, and lower associated risks. The technologies chosen for discussion in this book and the section in which they are presented are the following: Carbon Dioxide Injection (Section 2.2.1), Nitrogen Injection (Section 2.2.2), Polymer Flooding (Section 2.3.1), Alkaline-Surfactant-Polymer (ASP) Flooding (Section 2.3.2), Emulsion Injection (Section 2.3.3), Solvent-Only Injection (Section 2.4), Environmentally Friendly

Solvents (Section 2.4.2), Vapor Extraction Process (VAPEX) (Section 2.5.1), Subsurface Upgrading (SSU) via Cyclic Propane Injection (Section 2.5.2), Nsolv™ Technology (Section 2.5.3), Nanomaterials (Section 4.1), Ionic Liquids and Deep Eutectic Solvents (Section 4.2), Microbial-Enhanced Oil Recovery (MEOR) or Microbial-Derived Surfactants and Polymers (Section 4.3.1), Plant-Based Surfactants (Section 4.3.2), and Switchable Surfactants (Section 4.6). A complete discussion of each route can be found in its respective section.

1.5.2 Thermal Technologies

Thermal methods for HCO production involve the application of heat to reduce the viscosity of the oil, making it easier to extract from reservoirs in which non-thermal technologies are not effective or economically challenged. The most common thermal methods are hot water injection, cyclic steam injection, steamflooding, and SAGD. As depicted on the right-hand side of **Figure 1.5**, these technologies have been successfully implemented (TRL level 9, see **Table 1.2**) for the EOR of HCOs with multi-well and multi-field applications worldwide. However, due to their maturity and high degree of optimization, continuous improvement on these routes has a relatively low potential for reducing GHG emissions in the future (Lee 2021). Their main challenges are the considerable high energy and water consumption needed to generate steam (Smalley 2000, Gordon, 2015, and Rosa 2017). Furthermore, SSU via catalytic Aquathermolysis, *In-situ* Combustion-based technologies such as Toe-to-Heel Air Injection (THAI) and CAtalytic upgrading PRocess *In-Situ* (CAPRI) (middle of **Figure 1.5**) have been field-tested (Ovalles 2019), but due to their operating challenges and relatively large amount of GHG generated (Safaei 2019), their potential to reduce emissions is low. SSU using hydrogen donors has only been tested at laboratory conditions (Ovalles 2019).

As mentioned in the previous section, we selected the technologies depicted in **Figure 1.5** that have medium to high potential for GHG emission reduction vs. the Base Case (in this case, Steamflooding). This monograph will discuss these technologies in detail. We prioritized environmentally friendly approaches that are more likely to be deployed due to their technical scalability, lower cost, while also improving recovery factors for the HCO production. The technologies chosen for discussion in this book, are the following: Cyclic Steam Stimulation with Solvent Co-Injection (CCS-Solvent, Section 3.1.1), Steamflooding with Solvent Co-Injection (Section 3.1.2), Solvent/Steam-Assisted Gravity Drainage (SA-SAGD, Section 3.1.3), Steam/Additive Co-Injection (Section 3.1.4), Steam/Natural Gas Co-Injection Field Tests (Steam/NCG Field Tests, Section 3.2.1), Enhanced Modified Steam and Gas Push (eMSAGP, Section 3.2.2), Enhanced Modified Vapor Extraction (eMVAPEX, Section 3.2.3), Electromagnetic Heating (Radiofrequency) (Section 3.3), Solar Steam Generation (Section 3.4.1) Nuclear-Powered EOR (Section 3.4.2), Geothermal Energy/steam generation (Section 3.4.3), Nanotechnology and Nanomaterials (Section 4.1), Ionic Liquids and Deep Eutectic Solvents (Section 4.2), Ultrasound Applications (Section 4.4), and Subsurface Hydrogen Production (Section 4.5). A complete discussion of each route is available in its respective section.

1.5.3 Important Considerations

As mentioned, technology evaluation can be a complex task, as it involves diverse criteria, data availability, and different performance metrics. By using **Figures 1.4 and 1.5**, we conducted a high-level technology assessment based on their potential for field piloting and deployment as well as GHG emission reduction. However, it is essential to note that the technology comparisons illustrated in **Figures 1.4 and 1.5** primarily pertain to sandstone or clastic reservoirs, where most testing has occurred (Taber 1997, Alvarado 2010, and Vishnumolakala 2020). There is limited experience reported in carbonate reservoirs, so caution should be exercised when applying these results and conclusions to these environments.

It is essential to note that the technology used for HCO extraction should be tailored to the unique geographic and geological characteristics of each reservoir to ensure optimal production efficiency. Different countries adopt varying strategies (Seidy-Esfahlan 2024). For instance, Canada primarily employs bitumen mining, cold production with sand, and steam-based processes. Venezuela focuses on multilateral horizontal wells (cold production), electrical downhole heaters, and thermal methods. In the USA, Cyclic Steam Stimulation and steamflooding are quite common (Seidy-Esfahlan 2024). Therefore, when selecting technology for EOR of HCO, it is important to consider the specific reservoir properties, production techniques, and associated challenges.

Other technological risks were not considered in our technology assessment. Those are groundwater contamination with nanoparticles, ionic liquids, and other chemical additives, poor downhole mixing conditions, and the diffusion of solvents and other chemicals within the porous media. These topics will be discussed in Chapters 2–4.

Over the last few years, hybrid heavy oil EOR concepts have been reported in the literature that combines several technologies. For example, a consortium of several companies in Canada patented a process referred to as Enhanced Solvent Extraction Incorporating Electromagnetic Heating ("ESEIEH") for producing hydrocarbons from petroleum reservoirs. This technology involves pre-heating the mineral formation by using electromagnetic radiation (radiofrequency) and injecting a solvent in a SAGD well configuration (Trautman 2014). Also, the use of sonic energy or ultrasound has been a very useful tool in enhancing the reaction rates for a variety of chemical systems (Sawarkar 2009 and Avvaru 2018). Thus, it is not surprising that the use of ultrasound for the HCO production has been proposed in the literature. More details about both technologies are discussed in Sections 3.3.3 (ESEIEH Field Test) and 4.4 (Ultrasonic Applications). Hybrid concepts will be examined in detail in Section 5.2.

From the chemical characterization point of view, HCO has a relatively high carbon to hydrogen molar ratio (Altgelt 1994, Speight 2007 and Ovalles 2019, Chapter 2), and when burnt, they produce more carbon dioxide than conventional crude oils and natural gas. Thus, HCO could be better suited to be used as feedstocks to make petrochemicals or asphalt products than as fuels (Ovalles 2019). These conversion routes are not considered in this monograph.

REFERENCES

Ahmadi, M., 2023, *Sustainable In-Situ Heavy Oil and Bitumen Recovery.* Gulf Professional Publishing, Cambridge, MA, USA. 1st Edition | Elsevier Shop.

Altgelt, K.H., Boduszynski, M.M., 1994, *Composition and Analysis of Heavy Petroleum Fractions.* Marcel Dekker Inc., New York, p. 495. https://doi.org/10.1201/b16931

Alvarado, V., Manrique, E., 2010, "Enhanced Oil Recovery: An Update Review", *Energies*, 3, 1529–1575; https://doi.org/10.3390/en3091529

Atwater, I., Riva, J.P., 2020, "World Distribution of Heavy Oils and Tar Sands" Heavy oil and tar sand - World Distribution, Tar Sands, Energy Source|Britannica. Accessed February 18, 2025.

Avvaru, B., Venkateswaran, N., Uppara, P., Iyengar, S.B., Katti, S.S., 2018, "Current Knowledge and Potential Applications of Cavitation Technologies for the Petroleum Industry", *Ultrasonics Sonochem.*, 42, 493–507. https://doi.org/10.1016/j.ultsonch.2017.12.010

Beckman, M., 2018, "Cyclic Solvent Process Pilot: Non-Confidential Final Report Prepared For Emissions Reduction Alberta." Retrieved from https://www.eralberta.ca/wp-content/uploads/2019/01/CSP-Non-Confidential-Report.pdf. Retrieved March 6, 2025

Boduszynski, M.M., 2015, "Petroleum Molecular Composition Continuity Model" in "*Analytical Methods in Petroleum Upstream Applications*", C. Ovalles, and C. Rechsteiner Ed., CRC Press, Boca Raton, Chapter 1, and references therein. https://doi.org/10.1201/b18109

Bohm, M., Brasier, R., Keesom, B., Vogel, C., 2012, "A Greenhouse Gas Reduction Roadmap for Oil Sands." Retrieved March 3, 2025.

British Petroleum, 2016, "*BP Statistical Review of World Energy*" 65th Ed. London, United Kingdom, downloaded on August 22, 2018 from https://www.bp.com/content/dam/bp/business-sites/en/global/corporate/pdfs/energy-economics/statistical-review/BP-statistical-review-of-world-energy-2016-full-report.pdf

Canbaz, C.H., Temizel, C., Palabiyik, Y., Kor, K., Hendrandingrat, L., Bahar Hosgor, F., Aksahan, F., 2021, "Environmentally-Friendly Production and Recovery Processes for Heavy Presented at *SPE Trinidad and Tobago Section Energy Resources Conference Held Virtually* 28–30 June 2021. https://doi.org/10.2118/200897-MS

"Carbon Intensity Methodology," 2025, https://www.energyintel.com/wcod-carbon-intensity-methodology. Retrieved March 6, 2025

Department of Energy, 2015, "Technology Readiness Levels." Retrieved February 13, 2025.

DNV, 2025, "Energy Transition Outlook 2025" https://www.dnv.com/energy-transition-outlook/download/. Retrieved October 8, 2025.

Donaldson, E.C., 1979, "Environmental Aspects of Enhanced Oil Recovery", *Energy Sources*, 4(3), 213–229. https://doi.org/10.1080/00908317908908062

El-Houjeiri, H.M., Masnadi, M.S., Vafi, K., Duffy, J., Brandt, A.R., 2017, Oil Production Greenhouse Gas Emissions Estimator, OPGEE v2.0 User guide & Technical documentation. https://ww2.arb.ca.gov/sites/default/files/classic/fuels/lcfs/crude-oil/opgee_v2.0c_user_guide_technical_doc.pdf. Retrieved March 6, 2025.

Euhus, D., Sciamanna, S.F., Bryan, P.F., Wind, J.D., 2021, "Pathway to Process Technology Commercialization", *The Catalyst Review*, March, 6–12, The_Catalyst_Review_March_2021_condensed.pdf

Gomaa, S., Salem, K.G., El-Hoshoudy, A.N., 2024, "Enhanced Heavy and Extra Heavy Oil Recovery: Current Status and New Trends," *Petroleum*, 10 (3), September, 399–410 https://doi.org/10.1016/j.petlm.2023.10.001

Gordon, D., Brandt, A., Bergerson, J., Koomey, J., 2015, "Know Your Oil: Creating a Global Oil-Climate Index Carnegie Endowment for International Peace". https://assets.carnegieendowment.org/static/files/know_your_oil.pdf. Retrieved March 12, 2025.

Gray, M.R., 2015, *Upgrading Oilsands Bitumen and Heavy Oil*, The University of Alberta Press, Edmonton, Alberta, Canada. https://www.amazon.com/Upgrading-Oilsands-Bitumen-Heavy-Oil/dp/1772120359

Guzman, R., 2015, "Trends and Challenges in the Heavy Crude Oil Market" Arthur D. Little, 4th Heavy Oil Working Group, Bogota – September 22. Heavy and Extra Heavy Oil Forum, accessed February. 18, 20254

Huc, A., Argillier, J.-F., Renard, G., Quignard, A., Alazard, N., 2011, *Heavy Crude Oils; From Geology to Upgrading, an Overview*, Editions Technip, Paris.

IEA, 2025, "Oil 2025, Analysis and forecast to 2030", International Energy Agency, https://www.iea.org/reports/oil-2025. Retrieved November 6, 2025

International Energy Agency, 2024, "World Energy Outlook 2024," October, World Energy Outlook 2024 – Analysis - IEA accessed February 18, 2025

Islam, M.R., 2020, *Economically and Environmentally Sustainable Enhanced Oil Recovery*. Wiley, Hoboken, NJ, USA

Lee, J., Babadagli, T., 2021, "Mitigating greenhouse gas intensity through new generation techniques during heavy oil recovery," *J Clean Prod.*, 286, 124980. https://doi.org/10.1016/j.jclepro.2020.124980

Mai, A.A., Bryan, J., Goodarzi, N., Kantzas, A., 2009, "Insights Into Non-Thermal Recovery of Heavy Oil," *J. Can. Pet. Technol.*, 48(3), 27–35. Paper Number: PETSOC-09-03-27 https://doi.org/10.2118/09-03-27

Masnadi, M.S., El-Houjeiri, H.M., Schunack, D., Li, Y., Englander, J.G., Badahdah, A., Monfort, J.C., Anderson, J.E., Wallington, T.J., Bergerson, J.A., Gordon, D., Koomey, J., Przesmitzki, S., Azevedo, I.L., Bi, X.T., Duffy, J.E., Heath, G.A., Keoleian, G.A., McGlade, C., Meehan, D.N., Yeh, S., You, F., Wang, M., Brandt, A.R., 2018, "Global Carbon Intensity of Crude Oil Production," *Science*, 361(6405), 851–853. https://doi.org/10.1126/science.aar6859

McKinsey & Company, 2025, "Global Energy Perspective 2025" October. Retrieved October 14, 2025

Meyer, R.F., Attanasi, E.D., Freeman, P.A., 2007, "Heavy Oil and Natural Bitumen Resources in Geological Basins of the World", U.S. Geological Survey Open-File Report 2007-1084, available online at https://pubs.usgs.gov/of/2007/1084/OF2007-1084v1.pdf. Retrieved February 18, 2025

Morales, R., Rodriguez, F., Bejarano, C., Belhaj, H., Alhameli, F., AlDhuhoori, M., Quintero, E. 2024, "IOR/EOR Project Implementation Outcomes and Novel Insights in Extra-Heavy Oil Reservoirs in Eastern Venezuela Basin, Orinoco Oil Belt – Zuata Principal Field, Venezuela". Paper Presented at the ADIPEC, Abu Dhabi, UAE, November. https://doi.org/10.2118/222810-MS

Morrow, A.W., Mukhametshina, A., Aleksandrov, D., Hascakir, B., 2014, "Environmental Impact of Bitumen Extraction with Thermal Recovery", SPE No. 170066, presented at *SPE Heavy Oil Conference-Canada*, 10–12 June, Calgary, Alberta, Canada. https://doi.org/10.2118/170066-MS

Ovalles, C., 2019, *Subsurface Upgrading of Heavy Crude Oils and Bitumen*, CRC Press, Boca Raton: Taylor & Francis Group, https://doi.org/10.1201/9781315181028

Petersen, R., Sizer, N., Lee, P., 2014, *Tar Sands Threaten World's Largest Boreal Forest*, World Resources Institute, 15 July. Tar Sands Threaten World's Largest Boreal Forest | World Resources Institute. Retrieved March 18, 2025.

Ramirez-Corredores, M.M., 2017, *The Science and Technology of Unconventional Oils: Finding Refining Opportunities*, 1st Ed. Elsevier, London, Chapter 2, p. 41 and references therein.

Rosa, L., Davis, K.F., Rulli, M.C., D'Odorico, P., 2017, "Environmental consequences of oil production from oil sands, Earth's Future," 5, 158–170, https://doi.org/10.1002/2016EF000484

Safaei, M. Oni, A.O., Gemechu, E.D., Kumar, A., 2019, "Evaluation of Energy and Greenhouse Gas Emissions of Bitumen-Derived Transportation Fuels from Toe-To-Heel Air Injection Extraction Technology", *Fuel*, 256, 115930–115945. https://doi.org/10.1016/j.fuel.2019.115930

Sawarkar, A.N., Pandit, A.B., Shriniwas, D.S., Joshi, J.B., 2009, "Use of Ultrasound in Petroleum Residue Upgradation", *Can. J. Chem. Eng.*, 87, 329–342. cjce.2016920221214-1-1p9f82y-libre.pdf

Schenk C.J., Cook, T.A., Charpentier, R.R., Pollastro, R.M., Klett, T.R., Tennyson, M.E., Kirschbaum, M.A., Brownfield, M.E., Pitman, J.K., 2009, "An Estimate of Recoverable Heavy Oil Resources of the Orinoco Oil Belt, Venezuela", *U.S. Geological Survey Fact Sheet*, 2009–3028. https://pubs.usgs.gov/fs/2009/3028/, Retrieved December 17, 2016.

Seidy-Esfahlan, M., Tabatabaei-Nezhad, S.A., Khodapanah, E. 2024 "Comprehensive Review of Enhanced Oil Recovery Strategies for Heavy Oil and Bitumen Reservoirs in Various Countries: Global Perspectives, Challenges, and Solutions" *Heliyon*, 10, (18) e37826. https://doi.org/10.1016/j.heliyon.2024.e37826

Smalley, C., 2000, Heavy oil and Viscous Oil" *Modern Petroleum Technology*, R.A.D. Alan, ed., John Wiley and Sons Ltd., West Sussex, England, Volume 1, 6th Edition. Modern Petroleum Technology, 2 Volumes, Set, 6th Edition|Wiley

Speight, J.G., 2007, *The Chemistry and Technology of Petroleum*, 4th Ed., CRC Press, Boca Raton, USA, and references therein. https://doi.org/10.1201/9780824742119

Strausz, O.P., Lown, E.M. 2003, *The Chemistry of Alberta Oil Sands, Bitumen and Heavy Oils*, Alberta Energy Research Inst, Calgary, Alberta, Canada.

Taber, J.J., Martin, F., Seright, R. 1997, "EOR Screening Criteria Revisited-Part 1: Introduction to Screening Criteria and Enhanced Recovery Field Projects," *SPE Reservoir Engineering*, **12**(03) 189–198. https://doi.org/10.2118/35385-PA

Trautman, M., Ehresman, D., Edmunds, N., Taylor, G., Cimolai, M., 2014, "Effective solvent extraction system incorporating electromagnetic heating" US Patent No. 8,776,877.

UNITAR, 1982, "Proposal for a Definition of Heavy Crude and Sand", 2nd International Conference on Heavy Crude and Tar sands, United Nations Institute for Training and Research (UNITAR), Caracas, Venezuela, February 7–17, Chapter 1.

United Nations Climate Change, 2016, "The Paris Agreement" https://unfccc.int/process-and-meetings/the-paris-agreement, Retrieved December 3, 2025

Vishnumolakala, N., Zhang, J., Ismail, N.B. 2020, "A Comprehensive Review of Enhanced Oil Recovery Projects in Canada and Recommendations for Planning Successful Future EOR projects", SPE No. 199951. Presented at the *SPE Canada Heavy Oil Technical Conference held in Calgary*, Alberta, Canada, 29 September–2 October. https://doi.org/10.2118/199951-MS

2 Non-Thermal Processes

As mentioned in Section 1.5.1, non-thermal Enhanced Oil Recovery (EOR) technologies for extracting heavy crude oils (HCOs) do not add heat directly to the mineral formation. Instead, they utilize viscous forces and the internal drive energy of the reservoir (Mai 2009 and Ovalles 2019). These production methods (**see Figure 1.4**) have been successfully implemented (Technology Readiness level 9 [TRL level 9], see **Table 1.2**) for extracting HCOs due to their relative simplicity and cost-effectiveness. Typical recovery factors range from 5 to 12% of the Original Oil in Place (OOIP). However, there are several emerging technological approaches that are coming over the horizon that can have a significant impact on greenhouse gas (GHG) emission reduction and make those conventional non-thermal HCO production processes more environmentally friendly, while maintaining or improving the oil recovery factors. This chapter will describe recent developments in cold production, immiscible gas injection, chemical-Enhanced Oil Recovery (CEOR), solvent-only injection (hydrocarbon and environmentally friendly solvents), and subsurface solvent deasphalting. The potential GHG emission reductions and their respective TRL levels will also be examined.

2.1 RECENT DEVELOPMENTS IN COLD PRODUCTION

As shown in **Figure 1.4**, the conventional cold processes most commonly utilized in HCO production are cold production (without sand), Cold Heavy Oil Production with Sand (CHOPS), waterflooding, bitumen mining, and pressure pulse. This section presents detailed descriptions and recent developments of the cold production technologies. Other routes will be examined in other sections of this chapter, as shown in **Table 2.1**.

As mentioned in Section 1.5.3, different cold production technologies have been utilized depending on the unique geographic and geological characteristics of each HCO field. For example, in Canada, surface mining is effective for bitumen reservoirs located from the surface to 250 ft (75 m). The CHOPS method is suitable for heavy oil formations with thicknesses ranging from 1 m (3 ft) to 7 m (21 ft) consisting of unconsolidated sandstone. In Venezuela, primary production has been successful with the use of multilateral horizontal wells. In the USA, chemically EOR has shown promising results (Islam 2020).

2.1.1 Solution-Gas Drive and Multilateral Wells

As shown in **Figure 1.1**, HCOs and bitumen have a dead-oil viscosity at the original reservoir temperature between 100 and 10,000 mPa·s. These crude oils exhibit limited mobility under the temperature and pressure conditions found in reservoirs (Mai 2009 and Ovalles 2019). According to Darcy's Law, the oil can flow slowly

DOI: 10.1201/9781003635475-2

TABLE 2.1
Selected Conventional Cold Flow Technologies for Heavy Crude Oil (HCO) Production and the Sections in Which They Are Discussed in This Monograph

Conventional Cold Flow Technology	Discussed in Section
Waterflooding	Section 2.3. Chemical EOR Section 2.3.1. Polymer Flooding Section 2.3.2. Alkaline-Surfactant-Polymer (ASP) Flooding Section 2.3.3. Emulsion Injection
Cyclic or Continuous Solvent-Only Injection	Section 2.4. Solvent-Only Injection Section 2.4.1. Hydrocarbon Solvent Field Tests Section 2.4.2. Environmentally Friendly Solvents Section 2.5. Subsurface Solvent Deasphalting
Pressure pulse	Not considered (Seidy-Esfahlan 2024)

when high pressure gradients are applied. However, it has been observed that in these reservoirs, solution-gas drive results in significantly higher production rates and recoveries than conventional models of gas-oil relative permeability would predict (Yang 2019). This production mechanism, also known as foamy oil, was first identified in Canadian heavy oil and has since been observed in various reservoirs worldwide, including those in Venezuela, USA, China, and Russia (Mai 2009).

Considerable research has been performed in the past 35 years to understand the mechanics of foamy oil flow. The recovery from primary production in heavy oil reservoirs may be as high as 20% vs. the OOIP, but is usually lower (Mai 2009). It was found that the flow rates and oil recovery values are dependent on the existence of a high-pressure gradient across the core. **Figure 2.1** shows the typical behavior of the solution-gas drive mechanism (Islam 1990, Huerta 1996, Mirabal 1996, and Shen 1999). In general, the production occurs in saturated HCO reservoirs where the initial pore pressure exceeds the bubble point. As oil is extracted, the reservoir pressure decreases (light trace in **Figure 2.1**) and falls below the bubble point value. This situation causes gas to come out of solution and form bubbles within the oil phase. The expanding gas bubbles contribute reservoir drive energy to the system and help maintain reservoir pressure. Additionally, the presence of gas bubbles may reduce the viscosity of the oil, making it easier for the oil (black trace) to flow from the reservoir into the wellbore. Experimental and simulation results indicated that gas mobility in solution-gas drive in heavy oil is much less than in light oil, leading to improved recovery performance of the former (Pooladi-Darvish 1999). As expected, when the pressure decreases even further, production typically declines while gas production (dashed line) increases.

To optimize HCO production through the solution-gas drive mechanism, the use of multilateral horizontal wells has been successfully implemented in Venezuela, offshore Brazil, Nigeria, Ecuador, and various other regions worldwide (Gomaa 2024). For instance, in Venezuelan oil sands, the crude oil recovery factor can range

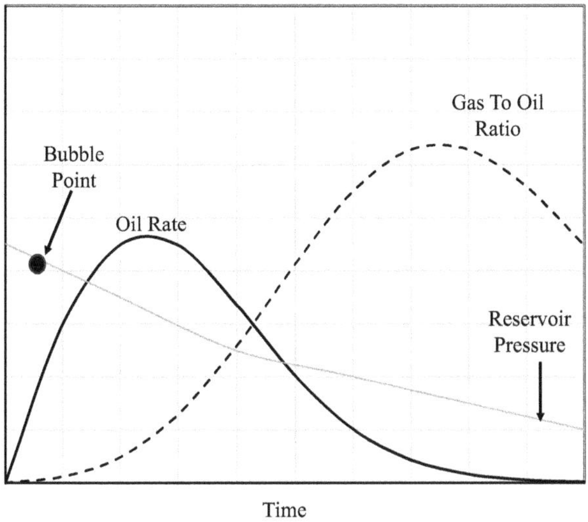

FIGURE 2.1 Oil rate, gas-to-oil ratio, and reservoir pressure typical profiles vs. time for the solution-gas drive mechanism.

from 4 to 12% of the OOIP (Morales 2024). Effective sand control, using slotted liners or other methods, is essential for the success of this technology (Seidy-Esfahlan 2024). Further enhancements can be achieved by incorporating artificial lift technology and cavity pumps. Additionally, applying diluents can help reduce crude oil viscosity and boost productivity (Gomaa 2024). This topic will be explored in Section 4.1.

Figure 2.2 depicts a lateral view of a multilateral horizontal well featuring both lined trunk and fishbone configurations. The oil-sand pay zone is initially identified using a pilot well on the left side of the **Figure 2.2**. Next, a deviated well is drilled to create the bottom horizontal section, known as the lined trunk, which extends away from the pilot well approximately 7,500 ft, as shown on the right side. Additionally, **Figure 2.2** illustrates a fishbone well designed to extract oil sands located above the lined trunk.

Several recent environmentally friendly technologies are emerging to help reduce GHG emissions, while maintaining or increasing oil rates. Bone et al. (2016) reported on various configurations of multilateral well technology that enhance the drainage area of reservoirs and increase oil production, resulting in significant economic benefits and lower GHG emissions. Some of the configurations discussed included multibranched wells such as dual, stacked, opposed, and forked laterals. Also, this approach reduces the number of wells needed and the spatial area required for well paths, thereby minimizing the environmental impact and the overall footprint on the oilfield (Bone, 2016).

Another technology that has the potential to reduce GHG emissions is Autonomous Inflow Control Devices (AICDs). These devices, including the Autonomous Inflow Control Valve, help manage fluid flow in wells by minimizing unwanted

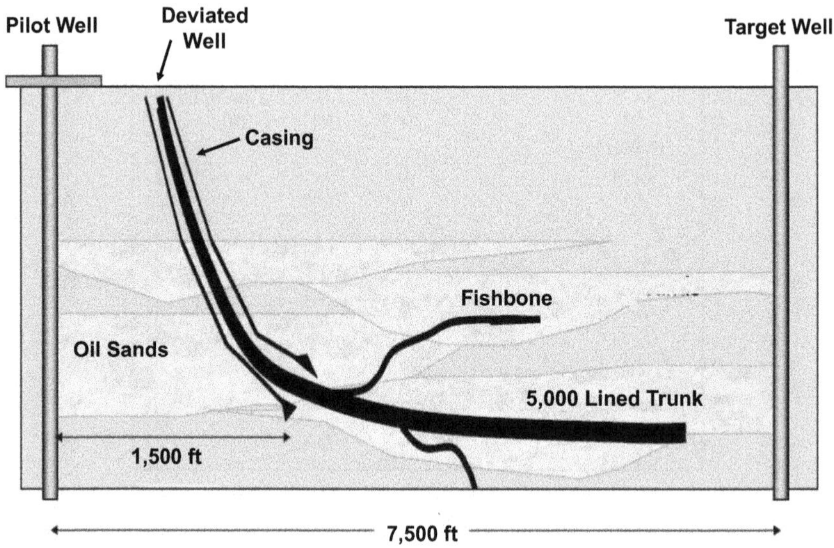

FIGURE 2.2 Lateral view of a multilateral horizontal well with lined trunk and fishbone configurations.

water and gas production while maximizing oil recovery. Moradi *et al.* reported a case study from a HCO field in the Marañon basin of the Peruvian Amazon, where a strong water drive was present (Moradi 2023). By employing AICD technology, they significantly reduced the volumes of water produced, leading to considerable savings in fluid lifting, treatment, and energy consumption associated with the operations in this field. The GHG footprint study indicated a reduction in carbon intensity of 56–78% when AICD completions were utilized compared to cases without AICDs (Moradi 2023). Similarly, Konopczynski *et al.* reported comparable results in an oilfield in Ecuador (Konopczynski 2024).

Almasri and coworkers optimized well flow and reduced costs by integrating multilateral wells, smart completion equipment, and AICD in a Middle East field (Almasri 2024). This approach enhanced well productivity, lowered construction costs, minimized environmental impact, and maintained optimal well flow while controlling water production.

Pinilla *et al.* compared the performance of AICD systems, inflow control devices, and slotted liner completions for the production of HCO from horizontal wells in the Llano Basin, Colombia (Pinilla 2025). The authors developed a three-dimensional, field-calibrated computational fluid dynamics model that demonstrated prediction deviations of less than 14% for oil rates and less than 10% for water rates. Replacing a slotted liner completion with optimally sized AICDs reduced cumulative water cut by up to 93%, lowered annular friction losses by 18%, and decreased CO_2 emissions by as much as 82%. The reduction in the carbon footprint was primarily attributed to a decrease in excessive water production (Pinilla 2025).

Based on the literature review presented, field tests, and deployments, multilateral well and AICD technologies have a TRL level 9 (multi-well and multi-field applications worldwide, see **Table 1.2**) and have high potential for reducing the GHG emissions, and at the same time, enhancing heavy oil recovery (see **Figure 1.4**).

2.1.2 Electrical Downhole Heater

A low-intensity thermal EOR technology, which we categorize as non-thermal because it mainly heats the wellbore, is the use of electrical downhole heaters (EDHs). These systems are specifically designed for heating the wellbore without adding energy directly to the reservoir by using a low-frequency electrical energy (50–60 Hz). So, this is "informally" a cold production technology even though energy is utilized. Technically, the heating is insufficient to create a steam chest like steamflooding or Steam-Assisted Gravity Drainage (SAGD). The primary objective is to reduce the viscosity of the HCO, thereby improving the flow from the lateral and toe sections of horizontal well producers (Shafiai 2020, Kumar 2023, Morales 2024, and Rodriguez 2024). Most of the heat generated remains within the producing tubing and is carried away by the produced fluids. Numerical simulations have indicated that the heat penetrates no more than 10 m (30 ft) into the surrounding formation, in the best-case scenario (Pizarro 1990, Ucan 2014, Quintero 2014, Medizade 2019, and Velasquez 2020). This technology is relatively straightforward to install and operate, and has proven effective when used in conjunction with submersible pumps.

Figure 2.3 illustrates a lateral view of a typical EDH installed in a horizontal production well. The well employs a progressive cavity pump (PCP), and the heater cable is powered from the surface using a standard power cable and an electrical transformer. The unit provides heat to the slotted liner through low-frequency

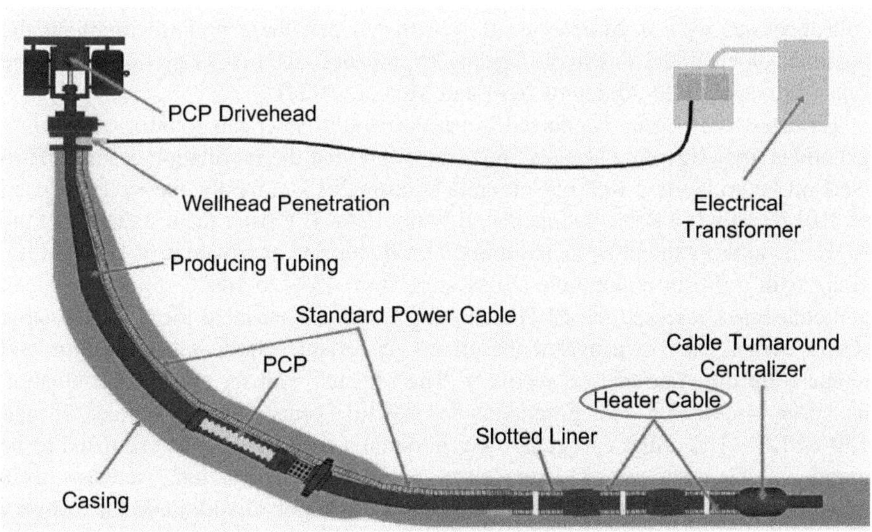

FIGURE 2.3 Lateral view of a typical electrical downhole heater (EDH) installed in a horizontal production well using a progressive cavity pump (PCP).

resistive heating. This type of DEH is most beneficial when the frictional pressure loss in the wellbore contributes back pressure to the reservoir sandface, and when the water cut is relatively low (less than 20%).

However, one drawback of this technology is that the installation of the heater cable can reduce the flow area within the wellbore. This reduction may limit the benefits of fluid heating on the wellbore pressure drop. Therefore, it is essential to carefully evaluate these two competing effects: the reduction in fluid viscosity vs. the reduction in wellbore area for each specific case.

Morales and colleagues conducted 23 installations of EDH systems in the Zuata Principal Field, located in the Eastern Venezuela Basin, between 2010 and 2018 (Morales 2024). They reported that a temperature of 160°F, measured at the discharge of the PCP, corresponded to a temperature gradient of approximately 40°F. This resulted in a significant viscosity reduction of about 60%. The observed oil flow rates ranged from 48 to 287 barrels per day (BBL/D), with an average of 100 BBL/D. These findings indicated an average production increase of 40% following the implementation of EDH, with variations ranging from 20 to 60%.

Quintero *et al.* conducted a history match between an EDH field test at the Zuata Principal Field and a numerical simulation model using Computer Modeling Group (CMG) STeam Additive Reservoir Simulator (STAR) CMG STARS (Quintero 2014). After successfully matching the production history before and after the implementation of electrical heating, they obtained values for the reservoir's thermal variables. Subsequently, they set up a second simulation model to forecast oil production over six years for a cluster of four wells. The results from this simulation were utilized to perform a techno-economic assessment (TEA) of the EDH technology within the Zuata field. The assessment revealed that the Net Present Value (NPV) was greater than zero, with an NPV of $12.5 million after 15 years of production and a payout period of two years (Quintero 2014). Both Quintero and colleagues, as well as Morales *et al.*, concluded that these findings highlight the potential of electrical downhole heating to enhance extra-HCO production in the Zuata Principal Field (Quintero 2014 and Morales 2024).

Field tests have been conducted in various South American countries, including Colombia and Argentina. Acosta and Garcés reported the installation of an EDH in the Don Pedro Norte 1 well in Colombia (Acosta 2013). Initially, the well produced 90 BBL/D with 0% Basic Sediment and Water (BS&W). After the installation of the EDH, the authors found an increase in oil production to approximately 180 BBL/D, along with a rise in bottomhole temperature from 43°C to 100°C. Similarly, Ucan and colleagues assessed the EDH technology in the Llancanelo Field in Argentina (Ucan 2014). They investigated the effects of variables such as heating time and temperature on additional oil recovery. The selected well for the EDH evaluation had been undergoing cold production for the first year, yielding between 38 and 150 BBL/D. The initial energy rate per horizontal well length was estimated to be around 40–60 watts per ft. Upon electrical heating, temperature increases were recorded, rising from 56°C to 100°C. The authors reported additional oil recovery ranging from 62 to 110% vs. the cold production. Importantly, no changes in reservoir temperature or viscosity were reported 30 m away from the wellbore (Ucan 2014).

Ojeda and Parman reported the trial-test results in Venezuela and California oilfields (Ojeda 2013). In the Californian case, the reservoir depth, pay zone thickness, temperatures, and American Petroleum Institute (API) gravities were 650–1,400 ft, 10–90 ft, 30°C–49°C (88–121°F), and 8.5°–21°, respectively. As in the previous cases, the main objective was to increase oil inflow in the pay zone by providing a substantial amount of heat into vertical wells via the electric downhole heaters. On average, for the 10 wells considered, higher temperatures up to 120°C (250°F) were observed, and oil production increased by 116%. Not all wells gave appreciably improved production, but a large majority did. In addition to enhanced oil production, improved pump life was observed, presumably due to reduced sand production and pump load. The authors concluded that EDH systems were very efficient, reliable, and easily controlled solutions to allow the production of heavy, high-viscosity oils (Ojeda 2013). The total system was very space efficient, both downhole and on the surface, which reduced the footprint, making it environmentally friendly.

Kumar *et al.* conducted an EDH trial test using a resistive-based unit in the Jodhpur Sandstone reservoir of Baghewala field of Western Rajasthan in India (Kumar 2023). This reservoir has an average depth of 1150 m, bottomhole temperature of 50°C, pressure of 11 MPa, and a pay zone thickness of 15–25 m. The produced crude has an API in the range of 14°–17° and viscosity from 8,000 to 15,000 cP @ 50°C. The EDH was installed along the walls of the production tubing in one of the horizontal wells, and its performance was compared to that of others that did not have a heater.

Kumar *et al.* found that oil production initially increased from less than 20 BBL/D to more than 80 BBL/D, eventually stabilizing at around 40–50 BBL/D after more than a year and a half of operation (Kumar 2023). They used a numerical model to show that the heat generated by the heater was confined to the wellbore and gradually transmitted to the surrounding formation. They suggested that this wellbore heating could help reduce or eliminate the skin damage caused by drilling and workover activities. Additionally, a decrease in water cut was observed, which corresponded with the rise in oil production (Kumar 2023).

A case study of an EDH is presented. The field trial was conducted in a long horizontal well (approximately 3,000–4,000 ft) to enhance the production at the Orinoco Oil Belt of Eastern Venezuela. The reservoir properties of this case study are reported in **Table 2.2**.

Figure 2.4 shows the results of the field test of the EDH. As seen, the temperature of the well increased from 48°C (119°F) at the toe to 57°C–62°C (134°F–144°F) at the heel, when the system turned on after 96 hours of operation. At the same time, the PCP speed increased from 150 RPM to 250 RPM, and the torque decreased by almost 13%. As expected, the oil production increase, reportedly from 980 BPD to 1176 BPD (~20%).

To determine the technical feasibility of a downhole heating process (Ovalles 2002), the energy gain (EG) was calculated by dividing the energy extracted as equivalent incremental oil (in comparison with the case when the heater is turned off) vs. the energy added into the reservoir when the downhole heating unit is turned on (see equation 2.1).

TABLE 2.2
Reservoir and Fluid Properties for the Field Trial of an Electrical Downhole Heater (EDH) at the Orinoco Oil Belt

Property	Value
Reservoir depth	730–914 m (2,400–3,000 ft)
Reservoir pressure	52–69 MPa (750–1,000 psi)
Pump intake pressure	24–48 mPa (350–700 psi)
Pay zone thickness	15–40 m (50–130 ft)
API crude (°API)	7.2–8.5
Gas-to-oil ratio	70–100
Viscosity	8,000–1,300 cP in temp. range of 43°C–60°C

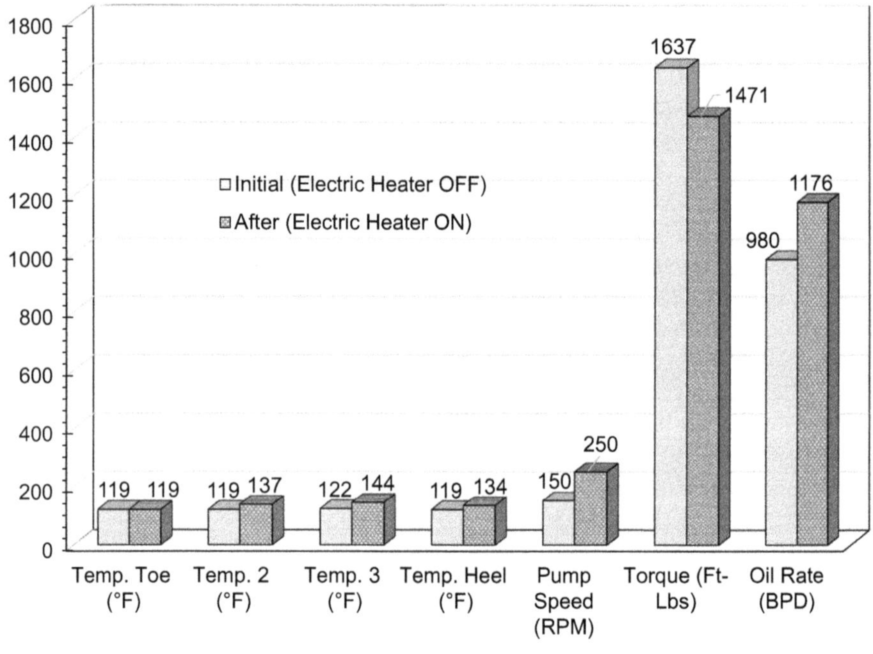

FIGURE 2.4 Results of the field test of the electrical downhole heater (EDH) at the Orinoco Oil Belt Reservoir.

$$EG = \frac{(\text{Increase in cumulative oil vs. Heater OFF}) \times 1{,}699\,\frac{\text{kWh}}{\text{BBL}} \times 0.5}{\text{Power used (kW)} \times \text{Days} \times 24\ \text{hours/day}} \tag{2.1}$$

In equation 2.1, a conversion factor of 1,699 kWh of one barrel of oil equivalent (BBL) was utilized, as reported in the literature (British 2022). Additionally, an efficiency rate of 50% was assumed for the processes of generating and transporting electricity to the remote areas of the Orinoco oilfield, reflecting a more realistic scenario. As indicated in equation 2.1, both the numerator and denominator are expressed in kWh of electricity, allowing for an estimation of the efficiency of the EOR process, that is, energy “out” divided by energy “in”.

In this case, the EDH had a rated power of 600 kW, and the test lasted for 96 hours. By applying equation 2.1, the EG calculated was 2.78×10^3 kWh/BBL. This positive value indicates that more energy is generated from incremental oil production than is consumed by the downhole heater. Moreover, if low-carbon-intensity electricity were employed, the increase in oil production could be achieved without relying on GHG-emitting steam generation technologies. Depending on the geographic location, a TEA is recommended to determine the feasibility of the project.

Based on the literature review, field tests, and commercial operations, the electrical downhole heating technology has a high TRL level 9 (multi-well and multi-field applications worldwide) and presents a medium potential for enhancing heavy oil recovery, and at the same time, reducing the greenhouse gas emissions, as shown in the technology assessment shown in **Figure 1.4**. An estimation of the carbon intensity is presented in Section 5.1.1.

2.1.3 Cold Heavy Oil Production with Sand (CHOPS)

This non-thermal cold production technology involves the deliberate and continued production of substantial quantities of sand along with the HCO. The sand influx creates cavities or “wormholes” near the wellbore, enhancing fluid flow and increasing oil production rates. **Figure 2.5** shows the lateral view of the CHOPS technology, and where the wormholes are located with respect to the vertical well. CHOPS is suitable for heavy oil formations with depths ranging from 400 m (1,300 ft) to 600 m (1,970 ft) and thicknesses from 1 m (3 ft) to 10 m (21 ft) (Seidy-Esfahlan 2024 and Society 2025a).

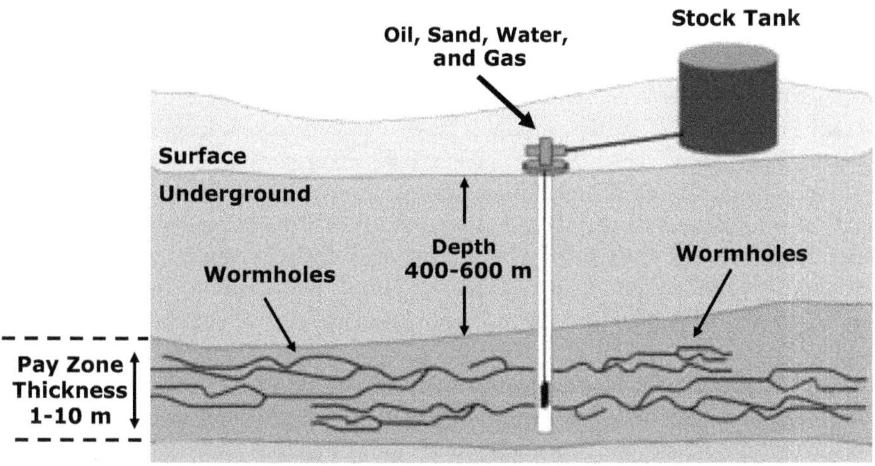

FIGURE 2.5 Lateral view of the Cold Heavy Oil Production with Sand (CHOPS) technology.

After the well is drilled into heavy oil reservoirs, oil production is initiated without sand-exclusion devices, such as screens or gravel packs. Sand production is encouraged during the completion phase to create cavities in the reservoir. As shown in **Figure 2.5**, oil, sand, water, and gas are pumped to the surface and stored in an insulated vertical separator (stock tank), where they are heated to temperatures between 60°C and 80°C. After allowing for sufficient residence time, the sand is separated and discarded. Oil production rates are approximately 30 m^3/day (190 BBL/D), and each well may produce between 10 and 20 m^3/day of sand during the initial days of production. This sand production typically diminishes to rates of 0.25 to 5 m^3/day once a steady state is achieved (Society of Petroleum Engineers (SPE) 2025b).

From an environmental point of view, produced sand contains 1–6% by weight of oil and is classified as non-hazardous oilfield waste (Society of Petroleum Engineers (SPE) 2025b). Other waste materials include produced water, which is typically cleaned and reinjected, as well as liquid wastes that contain varying amounts of oil and suspended fine-grained mineral matter, commonly referred to as "slops." Additionally, there are oil-in-water (o/w) emulsions, which consist of water, 20 to 50% oil (rich in polar asphaltenes), and fine-grained mineral matter. These emulsions are expensive to break down and present a significant challenge for the CHOPS industry (Society of Petroleum Engineers (SPE) 2025b).

CHOPS technology has been highly successful in Canadian reservoirs (Society of Petroleum Engineers (SPE) 2025a, Arnold 2012, Naderi 2007, and Speight 2016) and has also been implemented in various other regions around the world. Field tests and commercial production have been reported in Alaska (Bybee 2011), Sudan (Ruifeng 2011), Kuwait (Sanyal 2011), and China (Dusseault 2002). There are similarities in the reservoir rock and oil properties between the Canadian heavy oil areas and the Orinoco Faja fields, making CHOPS pilots a consideration for heavy oil reservoirs in Venezuela (Villarroel 2013).

In general, CHOPS recovery factors are in the range of 10–15% of the OOIP (Society 2025a). The environmental impact of CHOPS is significant due to the large footprint of the operating units, considerable GHG emissions from heated stock tanks (see **Figure 2.5**), and the challenges associated with handling and disposing of produced sand over the long term. In the past decades, innovative, environmentally friendly technologies have emerged to reduce sand handling, minimize lease sizes, and improve fluid storage and processing at the surface. These technologies are called Enhanced CHOPS. For instance, Arnold and coworkers reported the use of a SuperSump System, which aims to decrease the environmental footprint and boost recovery in typical CHOPS operations, with potential applications to both greenfield and brownfield heavy oil reservoirs (Arnold 2012). This technology produces oil from multiple wells in an upper reservoir into a specially designed cavern, which is created by washing salt formations beneath it. Inside the cavern, oil, water, and sand are separated, allowing clean oil to be produced through a single wellbore to the surface (Arnold 2012).

Arnold *et al.* conducted a simulation study on the Western Canadian Sedimentary Basin. Their results indicated that the SuperSump System could enhance recovery factors by up to 50%, lower operating costs by as much as 30%, and reduce GHG emissions by as much as 95% vs. conventional CHOPS (Arnold 2012). The authors estimated that the costs associated with this system represent a very favorable economic advantage compared to conventional CHOPS operations (Arnold 2012).

Wang and colleagues reported the implementation of integrated well completion strategies combined with CHOPS to enhance heavy oil production and minimize sand handling (Wang 2005). They limited the production of coarse sand while simultaneously extracting finer materials by utilizing a screen liner. Laboratory experiments demonstrated that sand particles measuring 0.9 mm could move to the surface alongside the heavy oil. Their study revealed that using a progressing cavity pump (PCP) initiated sand influx, but only 15% of the solids were allowed to create a wormhole that improved the permeability of the pay zone. The authors also presented initial results from field tests conducted on 20 heavy oil wells in the Fula oilfield, Sudan. The sand cuts in these wells ranged from 0.1 to 12% by volume, and the average oil production rates increased from 175 BBL/D to 580 BBL/D (Wang 2005).

Sonoro Energies announced the successful integration of CHOPS and a multilateral horizontal well in Kerrobert, Saskatchewan (Sonoro 2024). This combined approach increased the contact with the reservoir and has the potential to enhance oil recovery efficiency. The system was installed in March 2024, and the company reported no significant interruptions in production, achieving its highest production rate of approximately 80 BBL/D. Additionally, the well utilized its casing gas for energy, fuel, and heat, requiring minimal facilities and fewer overall multilateral wells compared to vertical and single-leg horizontal wells. This solution could offer long-term sustainability benefits by minimizing GHG emissions and reducing environmental impact (Sonoro 2024).

According to the literature review presented in this section, enhanced CHOPS technologies are at TRL levels between 5 and 6 (see **Table 1.2**), as indicated by the limited number of successful field pilot projects reported. Most findings are based on laboratory-scale studies, and the few field cases that exist show a moderate potential for reducing GHG emissions, as illustrated in the technology assessment (**Figure 1.4**).

2.2 IMMISCIBLE GAS INJECTION

Immiscible gas injection is an EOR technique in which gas or gas mixtures are injected into a reservoir at pressures below the Minimum Miscible Pressure (MMP), resulting in the oil and gas being mostly in separate phases (Society 2025c). While there is always a small amount of one phase dissolved in the other, we will set aside this phenomenon for the purpose of this discussion. The main reason for using immiscible conditions is when the MMP could not be achieved due to technical difficulties and/or economic considerations (Mahdavi 2018). Immiscible gas injection is particularly useful for HCOs and aims to increase oil production by reducing oil viscosity and increasing oil swelling (Society 2025c). In general, primary recovery processes can produce around 5–20% of the OOIP. The immiscible gas injection can increase the recovery for additional 15–20% of the OOIP, depending on the reservoir and fluid conditions (Zhang 2018).

In this section, carbon dioxide and nitrogen immiscible injections are presented. As discussed in Section 1.5.1, these technologies have been successfully field tested with limited adoption (TRL levels between 7 and 8, Demonstration) for HCO production (Chung 1988, Mahdavi 2018, Zhang 2018, Islam 2020, and Society 2025c) and have medium to high potential for reducing GHG emissions vs. the Base Case (cold production), as discussed during the technology assessment illustrated in **Figure 1.4**. It is important to mention that for any candidate oilfield, physical- and numerical-simulations-screening studies must be carried out to select (or "for selecting") the most suitable EOR method leading to maximum oil recovery.

2.2.1 Carbon Dioxide Injection

Immiscible CO_2 injection or CO_2 flooding has caught the public's attention in the last few years due to its cost-effectiveness and high displacement efficiency. In the USA, this method is particularly affordable because of the extensive natural CO_2 pipeline infrastructure. Moreover, CO_2 injection has the potential to reduce GHG emissions, making it an environmentally friendly EOR process (Mahdavi 2018 and Islam 2020).

The dominant occurring mechanisms for oil recovery during CO_2 flooding are through oil swelling, interfacial tension (IFT) reduction, pressure maintenance, and displacement toward production wells. Additionally, in conditions of immiscibility or partially miscible, there is considerable mass transfer of components between the oil and gas phases. However, immiscible or partially miscible CO_2 flooding is generally less efficient than miscible CO_2 injection and can be influenced by factors such as gravity segregation and microscopic fingering (Society 2025c).

The source of carbon dioxide is a key consideration in the development of CO_2 flooding projects. Typically, carbon dioxide is obtained from underground deposits and can also be captured from point source emitters such as power plants, refineries, cement manufacturers, and other industrial facilities (Zhang 2018). In the highly successful Permian Basin in Texas, the industry invested over $1 billion in a network of 2,200 miles of CO_2 transmission and distribution pipelines to support CO_2 flooding. Operators in the Permian Basin shared these costs across several large fields, which enhanced the economic viability of the CO_2 injection EOR processes (Islam 2020).

Zhang and coworkers analyzed 41 field cases reported in the open literature. Results showed that immiscible CO_2 injection could recover an additional 4.7–12.5% of OOIP with an average injection efficiency of 10.1×10^3 Standard Cubic Feet per Barrel (SCF/BBL) (Zhang 2018). **Figure 2.6** illustrates the minimum and maximum values for the reservoir properties observed in the immiscible CO_2 injection pilot tests. The findings suggest that immiscible CO_2 flooding is effective for light to HCOs, with °API gravities ranging from 10.8° to 39°API. The reservoirs studied had depths between 430 m (1,410 ft) and 2,590 m (8,500 ft), porosities in the range of 12–33%, and temperatures varying from 28°C (82°F) to 113°C(45°F). Other reservoir properties are shown in **Figure 2.6**. The authors concluded that, compared to CO_2 miscible flooding, immiscible CO_2 flooding shows significant potential across different reservoir and fluid conditions (Zhang 2018).

Based on the reservoir rock and fluid parameters, six different injection strategies have been utilized in immiscible CO_2 pilot tests. Those are water-alternating-gas

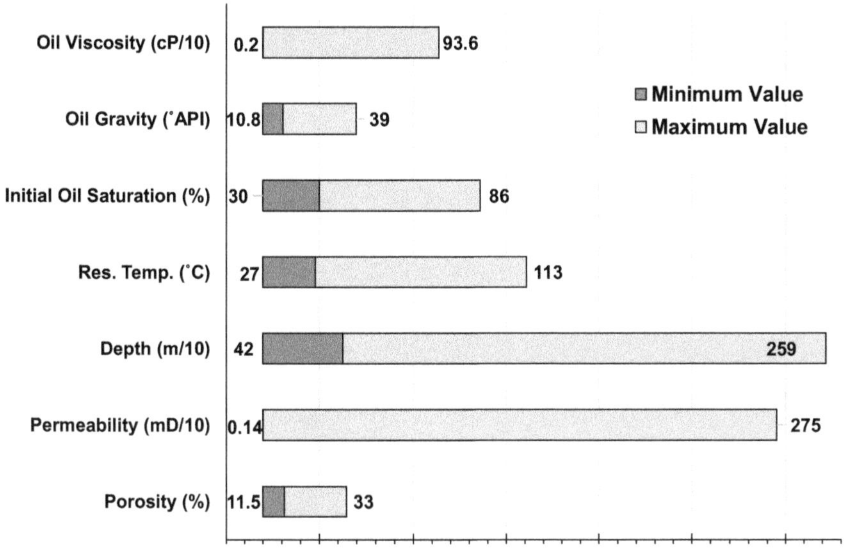

FIGURE 2.6 Minimum and maximum values for the reservoir properties for the immiscible CO_2 injection pilot tests reported in the literature.

Data taken from (Zhang 2018).

(WAG) injection, huff-n-puff, continuous CO_2 injection, simultaneous water and CO_2 injection, CO_2 slug process or intermittent injection, and carbonated water injection (Mahdavi 2018 and Zhang 2018). Among those, the first two are the most commonly used. WAG involves alternating the injection of water or oilfield brine with compressed CO_2. This approach utilizes water pressure to reduce gas channeling, which subsequently increases oil sweep efficiency. WAG is the most popular strategy, as it minimizes early CO_2 breakthrough and can lead to an average recovery increase of 9%. In contrast, the huff-n-puff method is valued for its simplicity and the wealth of experience in its application. In this process, carbon dioxide is injected into the production well for a duration of 6–60 days. After injection, the well is shut in for soaking for 10–13 days, followed by a production phase. Immiscible CO_2 injection has been applied in different parts of the world. Examples have been reported in Mexico (Abbaszadeh 2008), Iran (Fath 2014), India (Ganguli 2017), Middle East (Garnier 2023), Turkey (Gondiken 1987 and Sahin 2012), Italy (CORDIS 1985), and Indonesia (Panggabean 2022 and Ramadhan 2024).

Carbon dioxide injection projects are typically situated close to CO_2 sources to reduce transportation costs and enhance economic profitability. One notable case study is the Petra Nova Project. **Figure 2.7** shows the simplified process flow diagram of this initiative, the WA Parish CO_2-generating and capturing units (left-hand side), and the West Ranch CO_2-EOR facility on the right-hand side (Dubin 2017 and Islam 2020). Both parts of the project are located in East Texas, the first in Fort Bend County and the second in Jackson County. Petra Nova's post-combustion CO_2 capture system began operations in January 2017. The carbon-capture system is

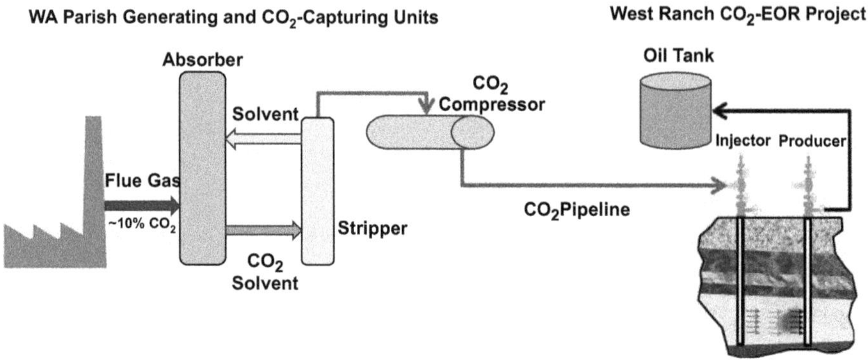

FIGURE 2.7 Simplified process flow diagram of the Petro Nova Project.

designed to capture about 90% of the carbon dioxide emitted from the flue gas or about 33% of the total emissions from the Unit (Dubin 2017). The project involved an ~80-mile CO_2 pipeline to connect the Parish CO_2-generating and capturing facility with the CO_2 injection wells (**Figure 2.7**).

Carbon dioxide was continuously injected using vertical wells at a depth of ~1,820 m (~6,000 ft) with a net pay of 9.75 m (32 ft), permeability of 500 milli Darcy, initial pressure of 9.3 MPa (2,800 psi), and temperature of 81°C (178°F). Until January 2020, the Petra Nova project produced over 4.2 million barrels of 30°API crude oil (equivalent to 11,500 BBL/D) at the West Ranch facility, and captured more than 3.54 million ton of CO_2 (approximately 10,000 ton/day) at the Parish CO_2-capturing units (Department 2020). The project's continuous operation for three years confirms the maturity of CO_2 capture, transportation, and injection processes and demonstrates their potential as environmentally friendly technologies for the future. Due to unfavorable economic conditions during the pandemic, the project was temporarily shut down in May 2020. On September 5, 2023, JX Nippon Oil & Gas Exploration Corporation announced the resumption of operations for the project until this date (Carbon 2023).

Another noteworthy case study is the immiscible CO_2 injection in Turkey's Bati Raman Field HCO reservoir (Sahin 2012). Since 1986, the Turkish Petroleum Corporation has been injecting approximately 1 million ton/day of naturally occurring CO_2 from the Dodan CO_2 gas field, located 55 miles away, into the Garzan limestone of the Bati Raman field. The reservoir fluid has an °API gravity of 9.7° to 15.2° and a viscosity ranging from 0.45 Pa·s (450 cP) to 1 Pa·s (1,000 cP) under reservoir conditions. As of December 2009, the recovery percentage had improved from 1.5 to 5.4% of the OOIP. A total of 323 billion SCF of CO_2 has been injected at a cost of $0.60 per 1,000 SCF (Sahin 2012). The Bati Raman immiscible CO_2 injection project is considered one of the most successful EOR initiatives in the extraction of heavy oil from fractured limestone reservoirs. Although a significant portion (~70%) of the injected carbon dioxide was back-produced, the project highlights the potential of this technology to significantly reduce GHG emissions.

Garnier *et al.* outlined a methodology for calculating GHG emissions from EOR technologies (Garnier 2023). The authors applied this workflow to a WAG-

CO_2 project in onshore oil reservoirs in the United Arab Emirates (UAE). They assessed the carbon dioxide balance and incremental CO_2 emissions arising from the additional facilities required for EOR, comparing these findings to a waterflood Base Case. The WAG-CO_2 project demonstrated the potential to store an average of 1.3 million ton of CO_2 in the reservoir while producing 83 million barrels of incremental oil. Due to the electrification of onshore fields in the UAE, the GHG emissions associated with this project were low. This preliminary evaluation indicated that nearly all imported CO_2 could be stored in the reservoir, creating a project focused on incremental oil recovery and net GHG emission reduction. To ensure the long-term storage of CO_2, it is recommended that a monitoring plan be established, along with the definition of post-EOR projects (Garnier 2023).

Abuov and colleagues conducted a life cycle assessment (LCA) of CO_2-EOR in major hydrocarbon basins around the world (Abuov 2022). They compared the estimated net GHG emissions of CO_2-EOR with those from the average oil produced in each respective country. When sourcing CO_2 from Natural Gas Combined Cycle (NGCC) power plants, most of the countries analyzed (the USA, China, Iraq, Saudi Arabia, Nigeria, Venezuela, Russia, and Kuwait) exhibited upstream-only GHG emissions of CO_2-EOR in the 180 ± 20 kilogram of carbon dioxide equivalent removed for one US Barrel of petroleum (kg CO_2e/BBL) range. According to the authors, this figure is lower than the average upstream GHG emission factor for the oil produced in those countries (~500 kg CO_2e/BBL) (Abuov 2022).

Morgan and coworkers (2023) studied the effect of a total GHG LCA on a multi-objective CO_2-WAG optimization with an integrated TEA, which factors in carbon tax credit. In the Base Case, 81% of the purchased CO_2 was sequestered in the Farnsworth field, Texas. However, the LCA results yielded a lower net estimate of 73% of the purchased carbon dioxide. Next, the optimized forecasted model, which took into account key operational and reservoir factors such as WAG ratio, injection rates and periods, and well specification resulted in an improved sequestration volume of 92% of the purchased CO_2. Nonetheless, when LCA was considered, the net value also dropped to 84%. These results clearly indicate a significant amount of net CO_2 is not accounted for when operations are not analyzed through LCA (Morgan 2023).

Haddad and Gates assessed the energy return and CO_2 sequestration in cyclic CO_2 and cyclic CO_2-hot water injection processes within a post-CHOPS heavy oilfield (Haddad, 2017). Their findings indicated that overall recovery factors can be improved through the appropriate design of the CO_2 follow-up process. Specifically, cyclic CO_2 injection results in an incremental recovery factor of 2.4% over four years of operation, along with a high energy return ratio. In contrast, CO_2-hot water processes can achieve higher recovery factors, but they come with lower energy return ratios. It is important to note that the amount of CO_2 sequestered in the reservoir is generally low, typically less than 5%. Therefore, these EOR processes are not strong candidates for effective CO_2 sequestration for post-CHOPS heavy oilfield exploitation (Haddad 2017).

According to the Clean Air Task Force, using one metric ton of anthropogenic CO_2 for EOR results in a reduction of 0.63 ton of GHG emissions during upstream operations. In other words, CO_2-EOR leads to a 37% decrease in CO_2 emissions per barrel compared to conventional oil production (Clean 2019). For HCO production, which has a carbon intensity of approximately 60 kg CO_2e/BBL (Masnadi 2018), the carbon intensity of using carbon dioxide for EOR is reduced to 37.8 kg CO_2e/BBL (Clean 2019).

It is crucial to note that the CO_2-EOR process has faced opposition from some environmentalists due to its link to increasing GHG emissions associated with crude oil production (Society 2025c). Additionally, Scope 3 GHG emissions have not been included in the TEA-LCA studies. Scope 3 are the direct and indirect GHGs emitted by the company sold products, that is, crude oil. This issue should be properly addressed in future studies.

Asphaltene precipitation and deposition are one of the main issues during miscible and immiscible CO_2 injection EOR processes. These issues can lead to pore plugging, reduced oil recovery, and damage to both surface and subsurface equipment (Leontaritis 1988, Hammami 1999, and Joshi 2001). Slimtube tests, long-core flooding, and nanopore shale samples have been used to simulate reservoir conditions to determine the operational envelope, gain mechanistic insights, and develop mitigation strategies (Elturki 2022 and Sofla 2024). The application of asphaltene stabilization agents can inhibit asphaltene precipitation during immiscible CO_2 injection, thereby reducing the risks of deposition and facilitating quicker restoration of well operations (de Boer 1995 and Gandomkar 2023).

Based on the previous literature review and pilot studies, it can be concluded that CO_2 injection technologies have been successfully field tested but with limited adoption (TRL levels between 7 and 8, Demonstration, see definitions in **Table 1.2**) for HCO production, and possess medium potential for GHG emission reduction (see **Figure 1.4**).

2.2.2 Nitrogen Injection

Immiscible nitrogen injection, also known as nitrogen flooding, involves injecting nitrogen gas (N_2) into an oil reservoir below the MMP. This process allows some nitrogen to partially dissolve in the oil, creating a vaporizing drive that helps increase the oil recovery factor (Orr 2007, Mahdavi 2018, and Tileuberdi 2023). Due to the relatively low pressure in HCO reservoirs, miscible nitrogen injection is generally not used because of the potential fracturing of the reservoir rocks and high costs associated with gas compression.

Under immiscible conditions, several challenges can arise during nitrogen flooding, including viscous fingering, channeling in high-permeability zones, and gravity segregation, which can lead to early gas breakthroughs. Immiscible nitrogen flooding is typically implemented in two ways: continuously injecting N_2 gas into a mature oilfield after waterflooding or using a WAG approach (Janssen 2018). **Figure 2.8** shows the lateral view of WAG under immiscible nitrogen injection. As shown, alternating nitrogen and water slugs are applied to control mobility and improve oil sweep efficiency. By periodically shifting between slugs of gas (N_2 or CO_2) and water, a gas breakthrough can be delayed, and petroleum contact time can be increased.

Janssen and coworkers compared various immiscible nitrogen injection methods, focusing on how rock-fluid and fluid-fluid interactions influence the immiscible flooding process (Janssen 2018). They performed core-flood experiments using Bentheimer sandstone cores, injecting nitrogen into cores that were saturated with

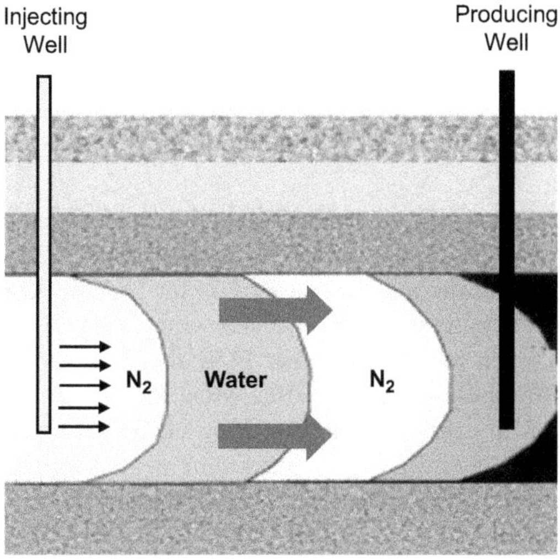

FIGURE 2.8 Lateral view of the water-alternating-gas (WAG) technology using immiscible nitrogen injection.

n-hexadecane at connate water saturation. The nitrogen was injected at constant pressures of 5 and 10 bar while varying the backpressure from 5 to 60 bar. Additionally, they injected nitrogen into residual oil during waterflooding and evaluated a WAG injection scheme. As with the case of the WAG-CO_2 process discussed in Section 2.2.1 (Mahdavi 2018 and Zhang 2018), Janssen's laboratory results showed that WAG-N_2 injection was more effective than continuous nitrogen injection. The authors concluded that the residual oil saturation during immiscible nitrogen flooding is lower in three-phase flow conditions compared to two-phase flow. The high oil recovery observed with the WAG method is likely due to the increased saturation of trapped gas (Janssen 2018).

Arias *et al.* evaluated a huff-n-puff cyclic solvent injection (CSI) technique as a non-thermal EOR alternative for increasing the heavy oil recovery in a shallow reservoir located in the middle Magdalena Basin in Colombia (Arias 2021). A pilot well with less than 30% water cut and 10.9°API was selected. Heavy natural gasoline of 30°API obtained from the same reservoir was injected using nitrogen as dispersing gas. Three cases were evaluated (Arias 2021):

1. Solvent slug injection of 60 BBL through the annular and dispersed by 46,444 m^3 (1,640 thousand standard cubic feet [MSCF]) of N_2 immiscible.
2. The same solvent volume as 1 was injected, but a third of the N_2 was pushed, that is, 15,481 m^3 (547 MSCF).
3. The same N2 volume as 2 was injected to determine the heavy oil response without solvent.

After soaking for 48 hours, the results indicated that the incremental oil productions, without accounting for the injected solvent, were 232, 120, and 11 BBL for cases 1, 2, and 3, respectively. As anticipated, the presence of solvent led to increased oil production with higher amounts of N_2 gas injected, due to the vaporizing drive of this gas. Additionally, it was found that immiscible N_2 injection alone was ineffective as a production mechanism without the influence of the solvent for this HCO (Arias 2021). Finally, caution should be taken to avoid causing asphaltene instability during N_2 injection, as reported by Jamaluddin and coworkers in their lab. experiments (Jamaluddin 2022).

The previous discussion highlights that several field pilot tests for immiscible N_2 injection have been successfully field tested but with limited adoption (TRL levels between 7 and 8, Demonstration, see **Table 1.2**) for HCO EOR (Tileuberdi 2023). However, there is currently no data available comparing the reduction of GHG emissions from nitrogen injection to other EOR methods. One clear advantage is that no GHGs will be back-produced as the cases for CO_2 and CH_4 flooding. The only GHG emissions to consider will come from the air separation unit involved in the process. Therefore, N_2-injection has medium potential for reducing GHG emissions, as illustrated during the technology assessment in **Figure 1.4**. This topic will be discussed in Section 5.1.

2.3 CHEMICAL EOR

CEOR generally involves the use of polymers, alkalis, and surfactants, as well as various combinations of these three types of chemicals. According to Alvarado and Manrique, CEOR experienced a peak in projects during the 1980s, but from 1990 to the present, there have been fewer initiatives (Alvarado 2010). This chapter will specifically focus on the most successful applications of CEOR for HCOs. As mentioned in the introductory chapter, HCO refers to petroleum oils with an API gravity of less than 20°API and a reservoir viscosity greater than 100 mPa·s.

The injection of non-indigenous chemicals downhole may introduce potentially hazardous material into the oil reservoir, and there is a possible risk of losing containment during workovers and production operations, which can cause damaging environmental effects. Most of those chemicals will not be brought back to the surface and will interact with the mineral formations in chemical pathways that are still under investigation (Nwosi-Anele 2019 and Islam 2020). Thus, in the last few years, emphasis has been placed on utilizing environmentally benign and/or sustainable materials and processes while obtaining higher oil recoveries and economic benefits.

Furthermore, recent technological advancements have significantly reduced both chemical-dosage requirements and CEOR costs, making the future outlook more promising (Mahdavi 2018, Islam 2020, and Rodriguez 2022). This section will introduce the chemicals involved (polymers, alkalis, and surfactants) and describe their fundamental mechanisms in HCO production. Case studies and the environmental impact of the technologies will be presented as well.

2.3.1 Polymer Flooding

The process of polymer flooding is the most widely used CEOR method and relatively easy to implement and operate (Saboorian-Jooybari 2016, Fassihi 2017, Rodriguez 2022, Goavec 2023, Garnier 2023, Quintero 2024, and Wei 2024).

While water quality requirements are not strict, low salinity is preferred. The handling and mixing of chemicals are simpler, although high-quality mixers should be used to prepare the source solution. Once the polymer is injected downhole, its distribution is typically less problematic compared to other CEOR technologies. Also, suppliers offer ready-to-go facilities made within sea transportation containers to easily deploy polymer facilities in the field. Percentages of recovery of up to 40% of the OOIP (typical values in the 20–30% range) have been reported (Delamaide 2014 and Delamaide 2025). For these reasons, in the past two decades, polymer flooding has been successfully implemented for the EOR of HCO worldwide.

Several studies have been conducted to understand the production mechanism (Fassihi 2017 and Delamaide 2025). Polymer flooding enhances macroscopic sweep efficiency due to its high viscosity, which facilitates the movement of oil and increases overall oil recovery. Recent studies indicate that, in addition to viscosity, the viscoelasticity of polymers can also be effectively employed to boost the recovery of HCO. This is achieved by mobilizing residual oil and improving microscopic displacement efficiency (Zeyna 2023).

The most commonly used polymer in CEOR is the polyacrylamide (PAM). **Equation 2.2** shows the hydrolysis of PAM into partially hydrolyzed polyacrylamide (HPAM). Depending on the reservoir conditions as well as the hardness and pH of the water, either PAM, HPAM, or their sodium salt analogs may be utilized for CEOR (Nagy 2025). PAM is primarily used as a viscosity enhancer in CEOR applications due to its molecular weight exceeding 1 million g/mol. HPAM is the most commonly used flow-modifier polymer because it can endure the high mechanical forces encountered during reservoir flooding. Additionally, HPAM is resistant to bacterial degradation and is also cost-effective. HPAM is not considered hazardous by the 2012 OSHA Hazard Communication Standard (29 CFR 1910.1200) (ThermoFischer Scientific 2025).

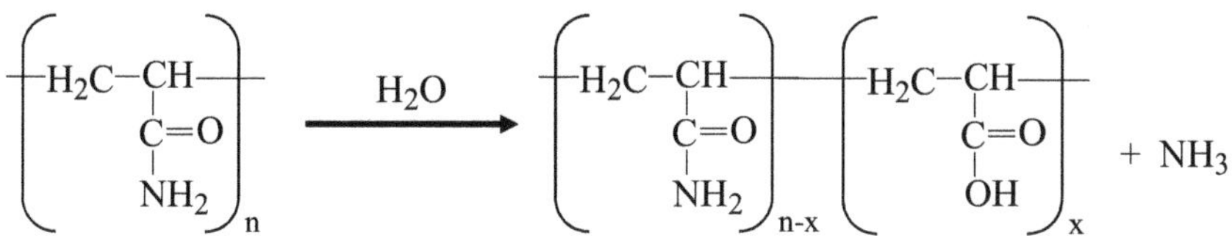

Polyacrylamide (PAM)

Partially hydrolyzed polyacrylamide (HPAM)

(2.2)

Almost 20 polymer CEOR project have been reported in the last 10 years in North America (Canada and USA), South America (Venezuela, Suriname, Colombia, and Argentina), India, Russia, and Oman (Saboorian-Jooybari 2016, Wei 2024, and Delamaide 2025). **Figure 2.9**. illustrates the minimum and maximum values for the reservoir properties for the polymer flooding pilot tests reported in the literature worldwide. As seen, polymer injection can be used when the reservoir viscosity is in the 27–415 cP range, with °API gravities ranging from 8° to 19°API. The reservoirs

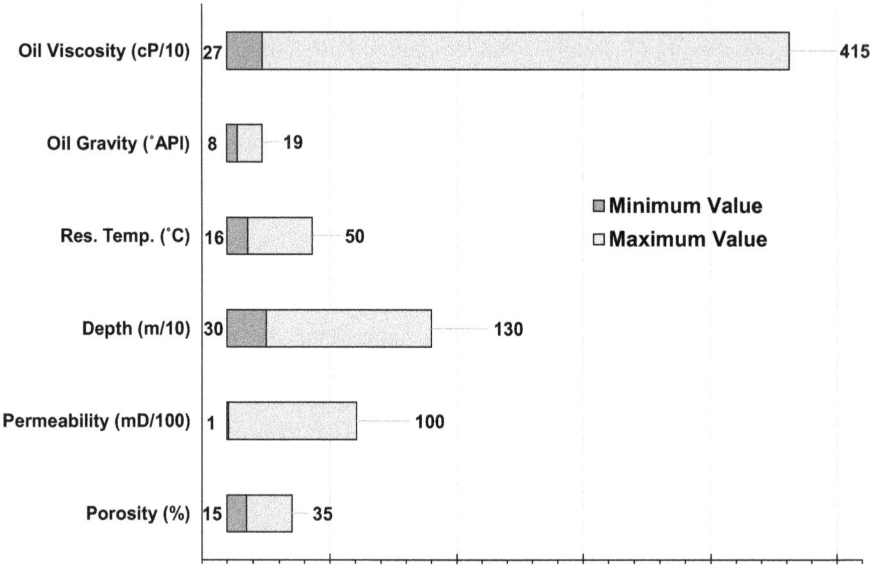

FIGURE 2.9 Minimum and maximum values for the reservoir properties for the polymer flooding pilot tests reported in the literature.

Data taken from (Saboorian-Jooybari 2016, Wei 2024 and Delamaide 2025).

studied had temperatures varying from 16°C (61°F) to 50°C (122°F), depths between 300 m (980 ft) and 1,300 m (4,260 ft), permeabilities from 0.1 to 10 Darcy, and porosities in the range of 15–35%.

Thermal EOR processes are generally the most widely used technologies to produce HCOs so it is useful to compare those routes with polymer flooding. **Table 2.3** shows that continuous steamflooding can be applied at depths of less than 1,500 m (4,900 ft). In contrast, polymer flooding is limited by the decomposition temperature of the polymer, which is generally below 140°C (284°F). Additionally, polymer injection can be effective in thin reservoirs (<18 ft), where thermal EOR methods are often considered unattractive or uneconomic due to significant heat losses.

Table 2.3 indicates that steamflooding can be applied to any crude oil or HCO viscosity; initial oil saturation greater than 40% and oil recoveries higher than 75% have been obtained. In contrast, polymer flooding is suitable for oils with a viscosity of less than 20,000 mPa·s (cP) and requires mobile oil, with recovery factors reaching up to 50%. Economically, **Table 2.3** shows that the capital expenses (CAPEX) associated with steamflooding are higher than those for polymer CEOR due to the need for high-temperature wells, pipelines, and larger facilities (footprint). In contrast, polymer flooding typically incurs lower operational expenditures and requires less complex water management compared to steamflooding. As seen, there are significant advantages in using polymer CEOR in comparison to continuous steam injection. Next, a case study at the Pelican Lake field in Canada will be presented.

TABLE 2.3
Comparison between Steamflooding and Polymer Flooding for Enhanced Oil Recovery (EOR) of Heavy Crude Oils (HCOs)

	Steamflooding	Polymer Flooding
Depth	< 1,500 m (4,900 ft)	Limited by temperature
Reservoir temperature	Any	<140°C (284°F)
Pay zone thickness	> 6 m (18 ft)	Any
HCO Viscosity	Any	<20,000 cP
Initial oil saturation	> 40%	Needs mobile oil
% Recovery factor (of the OOIP)	< 75%	< 50%
CAPEX	High	Low-medium
Aereal Footprint	Large	Low-medium
Operating Expenses (OPEX)	Large-medium depend on gas price	Low-medium
Water management	Large consumption require softening	Small consumption low salinity preferred
GHG Emission	Large	Low

The Pelican Lake field, located approximately 250 km north of Edmonton, Alberta, Canada, was discovered in 1978 and started producing in 1980. With more than 6 billion BBL of OOIP and a primary recovery estimated at less than 7%, it presents a significant target for EOR (Wilson 2015). However, it is also a challenging reservoir with high-viscosity oil in a thin formation. This sandstone reservoir has a shallow depth (450 m or 1,476 ft), high permeability between 0.8 and 8 Darcy, a thin pay zone (5 m [16 ft] on average), and an initial reservoir temperature of 16°C. The reservoir contains heavy oil (12°–14°API) with a viscosity that varies from 800 to 80,000 cP (Delamaide 2014).

Polymer flooding at Pelican Lake in Alberta began with a pilot project in 1997, although that initial attempt was unsuccessful. However, the concept was revisited and refined, leading to a successful second pilot in 2006 (Delamaide 2014a). It consisted of five horizontal wells, each 1,400-m long, including three production wells and two injection wells spaced 175 m apart. The noticeable response occurred between February and September 2006. The oil production rates increased almost one order of magnitude as follows: from 18 to 232 BBL/D in the first well, from 9 to 364 BBL/D in the central well, and from 16 to 139 BBL/D in the last well. The authors reported that these production rates remained high for nearly seven years (Delamaide 2014 and Wilson 2015).

The polymer flooding in the Milne Point field on the North Slope in Alaska is a very successful pilot and expansion, and is well documented in the literature (Aitkulov 2024 and Delamaide 2025). It consists of two reservoirs composed of shallow marine and fluvio-deltaic sediments with a thickness of 1–8 m (1–26 ft) and 6–12 m (20–39 ft), and good porosity (26–35%) as well as permeability (10–3,600 and 20–350 milli Darcy, respectively). The average crude oil viscosity is 850 mPa·s (cP).

During the first two years of testing, oil production rates doubled compared to the performance observed with waterflooding. It is projected that the ultimate recovery factor will be between 34 and 39% of the OOIP. These figures surpass the 19–21% recovery that can be achieved through waterflooding. Additionally, field observations indicated a "tailing" effect consistent with polymer retention, which enhances sweep efficiency (Delamaide 2025).

Another interesting case study is the Chichimene, a HCO field in the Colombian eastern basin (Quintero 2024). After successfully conducting a waterflooding EOR process since 2014, a polymer flooding pilot was carried out to increase the oil recovery. At the end of 2022, the incremental oil production had reached around 4.9 million BBL and is expected to rise to 6.9 million BBL, resulting in an additional ~10% incremental oil recovery factor in comparison with the waterflooding.

A carbon footprint, LCA, and energy intensity study of the polymer flooding was conducted (Quintero 2024). The main contributors to the GHG emissions and energy consumption were lifting, oil dehydration, water clarification treatment, and injection systems. The carbon footprint of the polymer injection project was calculated for eight years. The results showed a slight reduction in the carbon intensity vs. the baseline (range 20 to 19.3 kg CO_2e/BBL), whereas the energy consumption decreased to 6.7 kWh/BBL. This reduction was attributed principally to energy use optimization in crude oil lifting and enhanced water re-injection with polymer. The authors concluded that the carbon intensity reduction could avoid over 3,200 ton of CO_2e and improve the energy performance of polymer-EOR by more than 21% compared to waterflooding systems (Quintero 2024).

Skauge and coworkers compared the energy efficiency of waterflooding to polymer flooding to extract HCO from the Captain Field located in the North Sea (Skauge 2024). The oil viscosity is ~80 cP, and early water breakthrough in production wells, along with high-water cuts, were anticipated. While thermal EOR methods could address sweep efficiency concerns, these processes have considerable energy consumption and GHG emissions. Thus, a low-dosage polymer flood was implemented, yielding positive results. The authors observed that polymer flooding substantially accelerates the oil production, which results in a significant increase in NPV. The authors concluded that implementing low-dosage polymer flooding significantly enhances recovery rates, accelerating production by decades, and does so with reduced energy cost (Skauge 2024). Similar results were reported by Goavec et al. (2023) during the polymer injection in the Mangala field located in India. They found that polymer flooding is a more energy efficient process for oil recovery than waterflooding (Goavec 2023).

Based on the results presented in this section, it can be concluded that polymer flooding is an environmentally friendly alternative, offering a lower carbon footprint and greater energy efficiency compared to conventional waterflooding. Reports indicate that polymer flooding can help meet environmental targets by reducing CO_2 emissions per barrel by three to six times (Dupuis 2020). As shown during the technology assessment in **Figure 1.4**, it can be considered a commercial technology (TRL level 9), with multi-well and multi-field applications worldwide, and with high potential for GHG emission reduction.

2.3.2 Alkaline-Surfactant-Polymer (ASP) Flooding

ASP flooding is a well-established technology for enhancing the oil recovery of HCOs. It utilizes three types of chemicals, alkali (NaOH), surfactants, and polymers, to recover large amounts of residual oil not swept by waterflood. The main production mechanisms include decreasing IFT, increasing the capillary number, enhancing microscopic displacement efficiency, improving mobility ratios, and boosting macroscopic sweep efficiency (Mai-2009, Olajire 2014, Mahdavi-2018, Lee 2021, and Islam 2020). Incremental recovery factors for HCO applications are reported in the 20–40% range after waterflooding (Olajire 2014 and Nwosi-Anele 2019).

Several issues have been reported with ASP technologies. First, divalent cations (Ca^{2+}, Mg^{2+}) in clays and water react with alkali at medium to high pHs, which leads to the formation of silicate precipitates, resulting in scale problems and potential formation damage (Mahdavi 2018 and Lee 2021). Second, a relatively large amount of chemicals is consumed during ASP flooding, which, even at low dosage, could increase the operating costs. Third, surfactants and polymers may decompose at higher temperatures (>150°C–200°C) with the concomitant reduction in their efficiency. Lastly, alkalis and surfactants are known to form stable w/o emulsions, which can raise the operating and capital cost due to the required treatment on the produced fluids using surface facilities (Mahdavi 2018).

As mentioned, ASP is a commercially proven technology and several laboratory and pilot tests with light, medium and HCOs have been reported in the literature in North America (Sheng 2013), Canada (Delamaide 2014b), India (Dutta 2025), China (Olajire 2014), Venezuela (Rodriguez 2021), Colombia (Gutierrez 2024) and Nigeria (Nwosi-Anele 2019). An interesting case study is the Suffield Upper Manville ASP reported by Delamaide and coworkers (Delamaide 2014b). This sandstone reservoir has an average thickness of 2.9 m (9.5 ft), a permeability of 1–2 Darcy, and a temperature of 32°C. The API gravity ranges from 14° to 15°API, with a dead-oil viscosity between 480 and 600 mPa·s. The ASP project commenced in May 2007, utilizing an ASP solution consisting of 1–5% sodium hydroxide (NaOH), 0.1% alkyl aryl sulfonate, and 1,300 mg/L of Flopaan (a polymer). After primary production and waterflooding, the recovery factor was approximately 20% of the OOIP, with an incremental recovery of over 10% due to ASP flooding compared to the OOIP (Delamaide 2014b).

From an environmental perspective, non-thermal recovery methods, such as ASP flooding, do not require the generation of steam. This characteristic allows them to eliminate steam-associated CO_2 emissions entirely, making them one of the most environmentally friendly recovery methods available. Lee and Babadagli analyzed the carbon dioxide emissions associated with chemical EOR methods and found that water processing and facility electricity are the primary sources of GHG emissions (Lee 2021). Goavec and coworkers estimated reductions in the CO_2 emission intensity of the order of 30–40% in compared to waterflooding (Goavec 2023). As illustrated during the technology assessment in **Figure 1.4**, ASP chemical EOR technologies are commercially viable (TRL level 9) and have a medium potential for reducing GHG emissions.

2.3.3 Emulsion Injection

EOR through emulsion injection in heavy oil reservoirs involves injecting emulsions to improve oil recovery, particularly after primary production or waterflooding (Fassihi 2017 and Seidy-Esfahlan 2024). Binary emulsions can be classified into two types based on the dispersed phase: o/w, water-in-oil (w/o). They generally exhibit similar properties for EOR applications (Li 2024). These o/w or w/o emulsions can reduce water channeling, enhance sweep efficiency, and improve microscopic displacement efficiency with the concomitant increase of the displacement of residual oil.

A field test was conducted at the Section 5K of the Midway-Sunset Oil Field, California to evaluate the injection of o/w emulsions as an EOR technology (McAuliffe 1973). The sand thickness ranged between 3 and 6 m (ft), porosity from 33 to 35%, and oil saturation of approximately 40%. The emulsion was prepared using 70% Midway-Sunset Section 26 C crude oil (14°API) and 30% fresh water phase containing 1% NaOH. The average droplet diameter of the emulsion was 3 μm, and the viscosity was approximately 200 mPa·s. Over a two-year period following the emulsion treatment of three water-injection wells, fluid production from nearby wells exhibited increased oil recovery and a lower water-to-oil ratio. Additionally, the sweep efficiency was improved (McAuliffe 1973).

As with other chemical EOR methods, emulsion injection does not require steam generation so the steam-associated CO_2 emissions are not considered. Thus, emulsion flooding presents a promising solution for environmentally friendly 25–45 EOR technologies. Field tests have shown that emulsions can increase oil production by as much as 28% compared to traditional waterflooding methods (Fassihi 2017, Zhou 2019, and Seidy-Esfahlan 2024). The effectiveness of this technique largely depends on the characteristics of the emulsion and the specific properties of the reservoir. However, there have been no reported large-scale field or demonstration tests, which is why a TRL levels between 5 and 6 (see **Table 1.2**) has been assigned during the technology assessment (see **Figure 1.4**).

2.4 SOLVENT-ONLY INJECTION

In these processes, downhole solvent injection is conducted cyclically or continuously into the reservoir to dissolve HCOs, reducing their viscosity with a concomitant increase in oil recovery, without the need for thermal energy to be added to the formation (Lin 2014, Guo 2016, Bayestehparvin 2018, Ovalles 2019, and Ahmadi 2023). However, electrical energy may be required if the relatively expensive solvent is recycled. These EOR techniques have been studied and field-tested since the 1960s, utilizing both vertical and horizontal wells with varying degrees of success (Ahmadi 2023).

In the CSI technique, the process takes place within a single well, functioning as both an injector and a producer. When the solvent is injected into the reservoir, it diffuses throughout, helping to reduce the high viscosity of heavy crude oil and bitumen (HCO/B). After a soaking period of several days, the well is put into production, in which gravity directs the fluid with decreased viscosity to the producer.

CSI is similar to cyclic steam stimulation (CSS), but it does not involve adding thermal energy to the reservoir. CSI is also called "Solvent Huff 'n Puff." This CSI and production process continues for multiple cycles spanning months or years until the HCO produced becomes no longer economically viable. The cyclic operation is followed by a final blowdown period, during which additional solvent is recovered by vaporization at a low abandonment pressure (Canbaz 2021).

Generally, the solvents used in CSI include light aliphatic or aromatic hydrocarbons, as well as flue gas, low molecular weight ethers like dimethoxy ethane (DME), and mixed or unmixed form of CO_2 (Lim 1995, Firouz 2012, Lin 2014, and Canbaz 2021). Among the aliphatic hydrocarbons, ethane, propane, butane, cyclopentane, and paraffinic naphtha have been investigated. For the aromatic hydrocarbons, toluene, xylene, light oil cuts from refineries, and aromatic naphtha have been studied (Lin 2014, Canbaz 2021, Rodriguez 2022, and Ahmadi 2023). This section focuses specifically on hydrocarbon solvent injection, while CO_2 field testing (miscible and immiscible) is discussed in Section 2.1.1. The use of Cyclic Propane Injection for subsurface upgrading via deasphalting is examined in Section 2.5.2.

Alternatively, in a continuous operation, solvent injection is conducted in a SAGD well pair using short-chain paraffins, such as propane or butane, or mixtures thereof. The relatively high concentration of the hydrocarbon solvent leads to the *in-situ* asphaltene precipitation, which occurs simultaneously with the upgrading of HCO or bitumen (Ovalles 2019). This topic will be further discussed in Sections 2.5.1 (Vapor Extraction Process [VAPEX]) and 2.5.3 (Nsolv™ Technology).

During the CSI process, the composition of the heavy oil-solvent system changes, leading to mass transfer between the two components. Ahmadi conducted a review of both CSI physical and numerical simulations reported in the literature and concluded that the primary drawback of this method is the slow diffusion of the solvent into the HCO or bitumen (Ahmadi 2023). Laboratory, modeling, and field results indicate that heavy oil recovery increases due to the reduction of IFT, oil swelling effects, and a decrease in oil viscosity (Lin 2014).

The economic success of the SCI process heavily depends on effective solvent monitoring and recycling over months and years of operation (Lin 2014, Guo 2016, Bayestehparvin 2018, Ovalles 2019, and Ahmadi 2023). Typically, the cost per barrel of solvents is comparable to or more expensive than that of the HCO or bitumen produced. Furthermore, the presence of thief zones and fractures can lead to increased solvent retention in the reservoir, which in turn reduces the profitability of the CSI project. A potential source of low-cost solvent can be light oil produced from different reservoirs in the heavy oilfield. This approach was described by Garnett when light oil diluent from one reservoir was reinjected cyclically into the Monterey formation, which contained heavy crude (Garnett 2001). The issue of solvent high cost poses a significant challenge that must be properly addressed.

The CSI process has been considered for post-CHOP applications (Du 2013, Ivory 2010, Chang 2013, Chang 2015, and Lee-2021). In effect, Chang *et al.* (2009) numerically and experimentally simulated the CSI performance in horizontal wells as a follow-up process for CHOPS in the Cold Lake and Lloydminster reservoirs. After primary production, six cycles of 28% propane and 72% CO_2 were used to recover ~50% of OOIP. The numerical model matched well the experimental results

(Chang 2009). It has been proposed that wormholes provide reservoir access for injected solvent and a larger contact area for solvent to dissolve in oil (Chang 2015).

Four additional hybrids or modified CSI processes have been studied in the laboratory with various degrees of success. These concepts, as listed in the following, have the potential to reduce the environmental impact of HCO production:

1. Jamaloei *et al.* used propane as the solvent and methane as the chase gas during the solvent injection phase, with particular application to thin reservoirs. The authors reported percentages of oil recovery up to 40% (Jamaloei 2012).
2. Jia *et al.* 2013 studied pressure pulsing throughout the production phase by using a three-step pressure control methodology). Their findings revealed a 4-fold increase in the oil production rate and an 18% higher recovery factor compared to a typical CSI technique. The results were attributed to the formation of foamy oil as the primary mechanism for improving performance (Jia 2013).
3. Jiang and coworkers proposed the cyclic production with continuous vapor solvent injection as a source of additional drive for the diluted heavy oil (Jiang 2013). In addition, the producer is run in a shut-in/open cycle so that recovery factors of up to 85% could be obtained.
4. Sun and collaborators proposed a new process, called oil-soluble surfactant-assisted cyclic mixed solvent injection, to minimize solvent vaporization and sustain oil recovery during CSI (Sun 2019). During this process, an oil-soluble surfactant and a specially designed solvent mixture (produced gas and propane) are cyclically injected into the sand pack in two separate slugs. The most efficient lab test demonstrated oil recovery of up to 42% of the OOIP, compared to the traditional CSI process.

The potential to reduce GHG emissions of the CSI process has been studied by Soiket *et al.* (Soiket 2019) and Lee and Babadagli (2021). As anticipated, solvent-based technologies have lower energy requirements and, therefore, lower CO_2 emissions than their steam-based counterparts. Additionally, GHG emissions are mainly influenced by the solvent-to-oil ratio and how much of this material is lost in the reservoir vs. recycled. Additionally, energy consumption can be reduced by operating at a lower instantaneous steam-oil ratio (iSOR) in the case of steam-solvent co-injection (Soiket 2019). In the next sections, the hydrocarbon solvent field tests (Section 2.4.1) and the use of environmentally friendly solvents (Section 2.4.2) will be examined.

2.4.1 Hydrocarbon Solvents Field Tests

Several field test studies have been documented in the literature involving the injection of cyclic hydrocarbon solvents for HCO production (Lin 2014, Guo 2016, Bayestehparvin 2018, Ovalles 2019, and Ahmadi 2023). **Table 2.4** illustrates four case studies that serve as typical examples of this technology (Lechtenberg 1974, Chang 2015, JIVE 2010, and Beckman 2018). One of the earliest pilot tests

TABLE 2.4

Selected Field Tests Using Cyclic Hydrocarbon Solvent Injection for Heavy Crude Oil (HCO) Production Reported in the Literature[a]

S. No.	Field (Years)	Operator (Project)	Solvent	Injection Conditions[b]	Main Finding/Result[b]	References
1	Wilmington, California (1974)	US Bureau of Mines (NA)	Coker Oil (33°API)	4 vertical wells, 1 cycle 300–1,000 BBL 3–5 days soaking	• 37 and 36% increases in average oil rate and cumulative oil, respectively • On average, 5 BBL of additional oil was obtained for each BBL of solvent • ~75% of the solvent was recovered	Lechtenberg (1974)
2	Fruitvale, California (1974)	US Bureau of Mines (NA)	Residue (17.2°API)	One vertical well, 1 cycle, 300 BBL 3–5 days soaking	• 2–3-fold increase in oil rate • 80% solvent recovered	Lechtenberg (1974)
3	Lloydminster, Canada (2006–2010)	PTRC[c] and others (JIVE)[d]	Propane, butane, and CO_2	500,000 m^3 of Gas (propane/ CH_4) in horizontal wells	• Increased recovery by 29% in Post CHOPS wells • Reduce CO_2 emissions and water use by over 90% compared to SAGD	JIVE (2010), Chang (2015)
4	Cold Lake, Canada (2014–2018)	Imperial Oil Resources (CSP)[e]	Proprietary liquid solvent	Five CSI cycles in three horizontal wells	• CSI/CSP[d] technology has been deemed commercially viable • The GHG intensity reduction was 80% compared to CSS	Beckman (2018)

[a] NA = Not available or not reported.
[b] BBL= Barrels.
[c] PTRC = Petroleum Technology Research Center.
[d] JIVE = Joint Implementation of Vapor Extraction Project.
[e] CSP = Cyclic Solvent Project.

was conducted by Lechtenberg and colleagues from the US Bureau of Mines (Lechtenberg 1974). They performed the tests in five vertical wells located in two California oilfields, Wilmington and Fruitvale, which produced HCOs with an API gravity of approximately 15°API. These reservoirs are composed of unconsolidated sand with a bottomhole pressure of 1,000 psi (6.9 MPa) (Lechtenberg 1974).

As indicated in **Table 2.4**, entries 1 and 2, the field tests utilized highly aromatic solvents with API gravities of 33°API and 17.2° API (Lechtenberg 1974). The wells were allowed to soak for a duration of 3–5 days. For the Wilmington and Fruitvale fields, between 300 and 1,000 barrels of coker oil and 300 BBL of catalytic cracking residue were used per well. **Figure 2.10** illustrates the oil rates before and after one cycle of cyclic hydrocarbon solvent injection for the Wilmington field test. Notably, except for Well B, all other wells (A, C, and D) showed a significant increase in oil rates. The authors reported a 37% increase in average oil rate and a 36% increase in cumulative oil production, respectively. On average, five additional barrels of crude oil were obtained for each barrel of solvent used, with the percentage of solvent recovery of approximately 75% (Lechtenberg 1974).

For the Fruitvale Field test (**Table 2.4**, entry 2), a 2–3-fold increase in the oil rate and 80% solvent recovery after one cycle of CSI were reported (Lechtenberg 1974). The authors indicated that part of the solvent penetrated the reservoir, mixed with the crude oil, and increased the oil rates and recovery. However, they also suggested that the rise in oil production from the treated wells might be attributed to the removal of

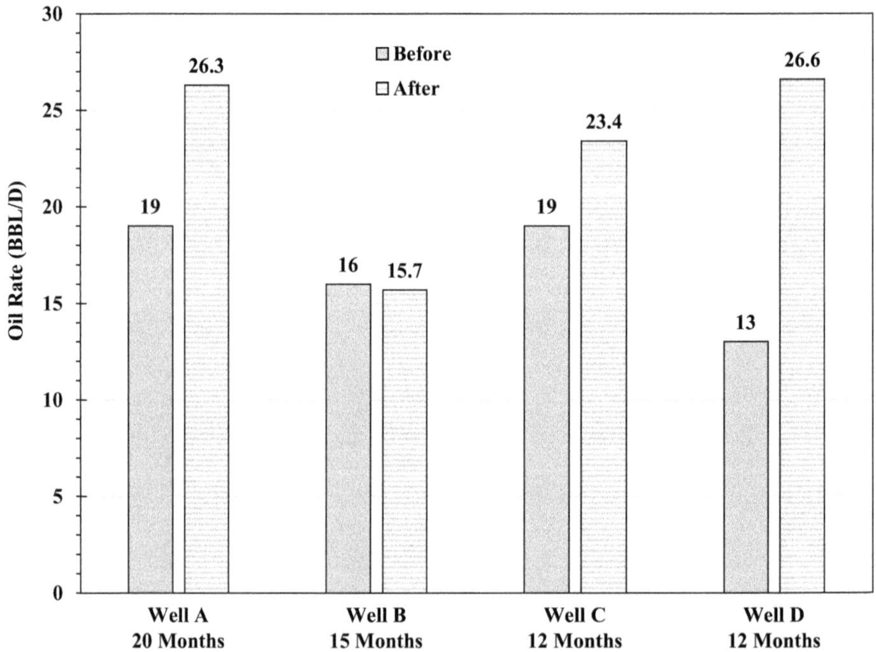

FIGURE 2.10 Oil rates before and after one cycle of cyclic hydrocarbon solvent injection for the Wilmington field test (Lechtenberg 1974).

asphaltenes and other solid deposits from the wellbore, casing perforations, gravel pack, and surrounding formation (Lechtenberg 1974).

The Petroleum Technology Research Center (PTRC), in partnership with Husky Energy, CNRL, Nexen, Alberta Innovates, and the Saskatchewan Research Council, conducted a $40 million Joint Implementation Vapor Extraction Project (JIVE) field test in the Lloydminster area of Canada from 2006 to 2010 (JIVE 2010 and Chang 2015). The solvents used were propane, butane, and CO_2, and they were injected in the gas phase (**Table 2.4**, entry 3). Two well configurations were tested in the JIVE program: one a horizontal injection and recovery method, and the other a cyclic horizontal injection and soaking program. The results indicated a 29% increase in oil recovery using post-CHOPS wells. Additionally, there was a reduction of over 90% in CO_2 emissions and water usage compared to conventional SAGD (JIVE 2010).

ExxonMobil, through its Canadian subsidiary Imperial Oil, field-tested (**Table 2.4**, entry 4) its patented Cyclic Solvent Project (CSP) in Cold Lake from 2014 to 2018 (Beckman 2018). In this process, a liquid-phase solvent is cyclically injected into a horizontal well. As discussed previously, the large mobility contrast between the solvent and the bitumen causes the solvent to finger into the bitumen, creating a mechanical dispersion and a large contact area for the rapid mixing of solvent into the bitumen. After injection, pressure is reduced, and the mixture of solvent and bitumen flows back to the same horizontal well, where it is produced to the surface using artificial lift. Cyclic injection and production operations continue for multiple cycles (>5) over several years until the bitumen produced is no longer economic. Three short horizontal wells were operated using CSP as a recovery process (Beckman 2018).

At high concentrations of propane, a mixture of propane and bitumen tends to separate into two distinct phases: a lighter liquid on top and a heavier liquid phase with a significantly higher density and viscosity at the bottom (Beckman 2018). A similar phenomenon was observed by Ovalles *et al.* in their pressure-volume-temperature (PVT) studies of propane and HCO conducted at 50°C and 7.49 MPa over a period of five days (Ovalles 2017). After the CSP field test encountered plugging issues during the early stages of operation, the diluent used was replaced with a proprietary flow-assurance solvent to minimize the formation of heavy liquids under pilot conditions (Beckman 2018). The author concluded that the CSP field test met its overall objectives, demonstrating that the technology is commercially viable. Furthermore, the GHG intensity reduction was reported to be ~78% compared to CSS (23.7 kg CO_2e/BBL for CSP vs. 106.7 kg CO_2e/BBL for CSS), which could potentially broaden its applications in other oilfields (Beckman 2018).

Based on the previous literature review and pilot studies, it can be concluded that hydrocarbon solvent injection technologies have been successfully field-tested but have seen limited adoption (TRL levels between 7 and 8, Demonstration, see **Table 1.2**) for HCO production. These CSI technologies show medium potential for reducing GHG emissions (see **Figure 1.4**).

2.4.2 Environmentally Friendly Solvents

As mentioned above, conventional solvents used in CSI typically include light aliphatic hydrocarbons, such as propane, butane, cyclopentane, and paraffinic naphtha,

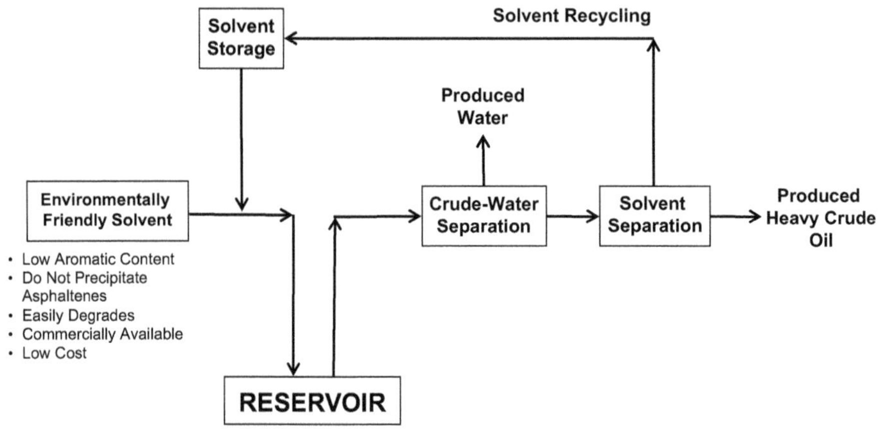

FIGURE 2.11 Schematic diagram for the Enhanced Oil Recovery (EOR) of HCOs and bitumen utilizing environmentally friendly solvents.

as well as aromatic hydrocarbons like toluene, xylene, light oil cuts from refineries, and aromatic naphtha (Lin 2014, Canbaz 2021, Rodriguez 2022, and Ahmadi 2023). However, these materials can be relatively toxic (Fischer 2013), which poses significant challenges for their handling and transportation to and from oilfields, particularly in remote and environmentally sensitive areas (Canbaz 2021 and Lin 2014). Furthermore, if there is a subsurface loss of containment during the CSI process, these solvents could result in groundwater contamination and potentially hazardous surface expressions. Therefore, the use of environmentally benign solvents is suggested as an alternative for wellbore and formation cleaning (Chandler 1996 and Berry 2007), and for EOR of HCOs (Espinosa 2022, Woolford 2022, and Mathews 2022 and 2024).

Figure 2.11 presents the schematic diagram for the EOR of HCOs and bitumen utilizing environmentally friendly solvents. These diluents are selected so that they have a low aromatic content, thus low toxicity, and do not lead to wellbore obstructions and formation damage via asphaltene precipitation (Woolford 2022). In the event of loss of subsurface containment, it is essential that these solvents can chemically degrade into ecologically compatible products, resulting in minimal or no long-term environmental effects. Finally, these environmentally benign solvents should be commercially available at relatively low cost to increase the economic feasibility of the CSI process. Some of the benign solvent options will be addressed in the ensuing paragraphs.

As discussed, CSI is a method that utilizes a solvent to dissolve HCO, thereby reducing its viscosity and enhancing its mobility and production rates. Espinosa *et al.* reported that the addition of 5% D-limonene decreased the viscosity of a HCO from 117 Pa·s (117,000 cP) to 55 Pa·s (55,000 cP) at 25°C, which corresponds to a 53% reduction (Espinosa 2022). Similarly, Woolford and coworkers found that 10% D-Limone reduced the viscosity of the blend from 469 cSt to 97.1 cSt, representing a 79% reduction (Woolford 2022). As shown in **Figure 2.11**, the CSI process can be conducted continuously or cyclically. After water separation, the environmentally benign solvent can be recycled, further enhancing the economic prospects of the process. Essential oils retrieved from citrus fruit peels, particularly the more abundant and cheaper orange oil, are the most important natural sources of D-limonene. In addition, tire pyrolysis oils provide large quantities of this compound as a by-product from Diels-Alder addition of isoprene.

The most environmentally friendly solvents studied in the literature are natural-sourced terpenes (Berry 2007, Espinosa 2022, Woolford 2022, Mathews 2022 and 2024). These naturally occurring organic compounds are found in many plants, particularly trees and citrus fruits. They are safe to handle and transport, and are commonly found in citrus- and pine-based soaps, cleansers, detergents, and fragrances (Fischer 2013). Terpenes possess solvency properties comparable to those of aromatic diluents, such as xylene and toluene, making them effective in dissolving waxes, asphaltenes, and paraffins (Curtis 2003). Furthermore, these products have a higher flash point than xylene, and, thus, are safer to handle. They are sustainable, renewable, and biodegradable (Fischer 2013 and Matthews 2024). Examples of terpenes are D-limonene, beta-pinene, myrcene, terpinolene, alpha-farnesene, beta-caryophyllene, nerol, citral, camphor, menthol, geraniol, camphene, squalene, humulene, carvone, linalool (Woolford 2022).

Woolford and coworkers from Chevron conducted numerical simulations of CSI in four vertical wells using pinene as a solvent for a typical California reservoir (Woolford 2022). **Figure 2.12**. shows the modeling results utilizing 2,000 BBL of Pinene per slug and six injection cycles between years two and five of the study. The results indicate significant increases in oil rates and cumulative oil production when

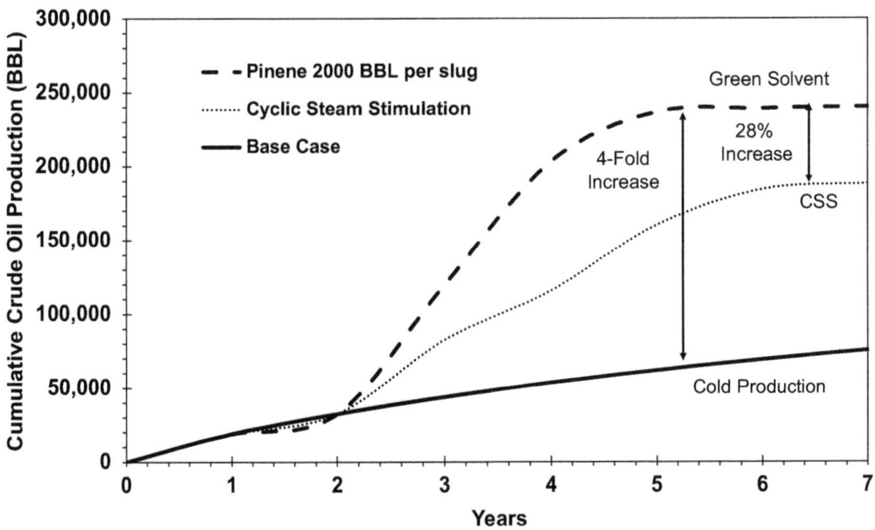

FIGURE 2.12 Numerical simulations of CSI in four vertical wells using pinene as solvent for a typical California reservoir.

Data taken from (Woolford 2022).

compared to the Base Case (cold production) or CSS. Specifically, the average oil rate increased from approximately 10 BBL/D to over 150 BBL/D following the solvent injection during the seven-year project. As shown in **Figure 2.12**, there was a 4-fold increase in cumulative oil production for the cyclic pinene injection compared to the Base case and a 28% increase vs. CSS (Woolford 2022). Similar outcomes were observed using D-limonene, although those results are not shown here.

Mathews and coworkers from Texas A&M University explored the potential of four terpenes (D-limonene, turpentine, citronella, and beta-pinene) as green solvents for enhancing heavy oil recovery. They conducted a series of core flooding experiments to compare the effectiveness of these environmentally friendly solvents with that of toluene (Mathews 2022 and 2024). The results indicated that the terpenes achieved 75% oil recovery, with efficiencies comparable to those of toluene. Additionally, all the terpene solvents acted as demulsifiers, which improved the dewatering of oil produced during the lab experiments. Notably, the co-injection of carbon dioxide with D-limonene further EOR efficiency by 2-fold, highlighting the potential of terpene-based solvents in environmentally conscious heavy oil recovery practices (Mathews 2024).

BioOilSolv is developing a biogenic solvent derived from renewable agroforestry biomass to improve bitumen production in Solvent-Assisted SAGD operations and simultaneously reduce GHG emissions (Advanced 2025). This Canadian-based pilot project, with a budget of Can$12 million, aims to demonstrate the feasibility of producing the solvent on a pilot scale, optimize its use in SAGD recovery, and validate its performance through both laboratory and field-scale testing. The successful completion of this project is expected to achieve a 35% reduction in GHG emissions, lower the steam-to-oil ratio (SOR) in SAGD operations, and increase bitumen production rates. No timeline has been provided for the start of the pilot project (BioOilSolv 2025).

Initial high-level TEAs and LCAs have indicated that CSI using environmentally friendly solvents has a significant potential to reduce GHG emissions compared to traditional steam-based methods (Lee 2021). However, due to seasonal variability and cost fluctuations associated with plant-derived solvents, more comprehensive TEA and LCA studies are needed to provide a thorough evaluation. Additionally, because these "green" solvents are plant-based, the carbon intensity can be speculated to be in the 10–30 kg CO_2e/BBL range. However, no field test has been conducted up to the publication of this monograph. Therefore, a TRL levels between 3 and 4 (lab/numerical simulation, see **Table 1.2**) can be assigned as illustrated in **Figure 1.4**. More details will be presented in Section 5.1.

2.5 SUBSURFACE SOLVENT DEASPHALTING

In these technologies, a hydrocarbon solvent is injected downhole, resulting in asphaltene precipitation through solvent deasphalting (SDA). The objectives are to increase oil rates and percentages of recovery, and at the same time, permanently improve the properties of the produced oil, such as API gravity, viscosity, amount of distillable material, sulfur, and metal contents. (Ovalles 2019). A key advantage of SDA processes is that they do not require reservoir heating, as they operate at reservoir temperatures (Pourabdollah 2013) or with warm solvents (Nenninger 2005).

It is crucial to mention that asphaltenes are a solubility class of compounds and are arbitrarily defined as the fraction of petroleum that is insoluble in light alkanes (C3-C7) but soluble in aromatic (benzene, toluene, xylene, etc.) solvents (Ovalles 2019). Asphaltenes play a critical role in the rheological behavior of HCOs as a result of their particular physical and self-associating properties (Ovalles 2019). By removing the asphaltenes from the heavy crude, reductions of two to three orders of magnitude on the viscosity are obtained over all the entire temperature range (Ovalles 2012).

The physical and numerical simulations of subsurface solvent deasphalting and asphaltene precipitation have been thoroughly reviewed (Ovalles 2019). In this book, we will focus on several field tests to demonstrate their EOR performances as well as their effectiveness as environmentally friendly technologies. Those processes are VAPEX, Cyclic Propane Injection, and Nsolv™ Technology.

2.5.1 Vapor Extraction Process (VAPEX)

The VAPEX process involves injecting solvent vapor downhole, typically consisting of light hydrocarbons in the C3 to C5 range, into an upper horizontal well within a conventional SAGD (Steam-Assisted Gravity Drainage) configuration pair (see **Figure 2.13**) (Butler 2000, Das 1998, Das 2005, and Ahmadi 2023). As reported in the literature (Ovalles 2019), the hydrocarbon solvent dissolves into the viscous oil, enhancing its mobility and allowing it to flow down to the production well. Laboratory experiments have demonstrated that asphaltene precipitates occur *in-situ*, leading to a slight improvement in oil density from 6° to 7° API.

The VAPEX process was field-tested at the Dovap site in Canada. This project involved a Can$30 million, well-instrumented, seven-year pilot study. The design included one cold-startup well pair and one hot-startup well pair. **Figure 2.13** illustrates the well array and surface process diagram for the VAPEX field test. As shown, the surface facilities were equipped to inject a methane/propane gas mixture downhole and to separate the produced fluids into gas (which was sent to flare) and solvent-free oil. Steam facilities were also included in the design.

The cold-startup well pair experienced communication issues, which hindered its ability to deliver the expected results. From September 2005 to July 2006, the oil production rate was less than 100 BBL/D (~15 m^3/day), despite an upgrade from 10.5°API to 16.8°API. During this period, reductions in asphaltene content were observed, decreasing from 17 to 1 wt.%, along with a decrease in viscosity. In contrast, the hot-startup well pair performed as anticipated, maintaining temperatures around 300°F from toe to heel, and achieving oil production of up to 100 m^3/day with a SOR of approximately 2.6. Unfortunately, the transition from SAGD to VAPEX was unsuccessful, leading to a decline in oil production to less than 20 m^3/day. Similar to the cold startup, this well pair also saw an upgrade from 10.5°API to 12°API, with reductions in asphaltene levels from 17 to 5 wt.%, as well as a decrease in viscosity. The project was ultimately discontinued in 2008 (Pourabdollah 2013).

Based on the previous discussion, the VAPEX process has been field tested with mixed results. In terms of environmental impact, it can be argued that the VAPEX

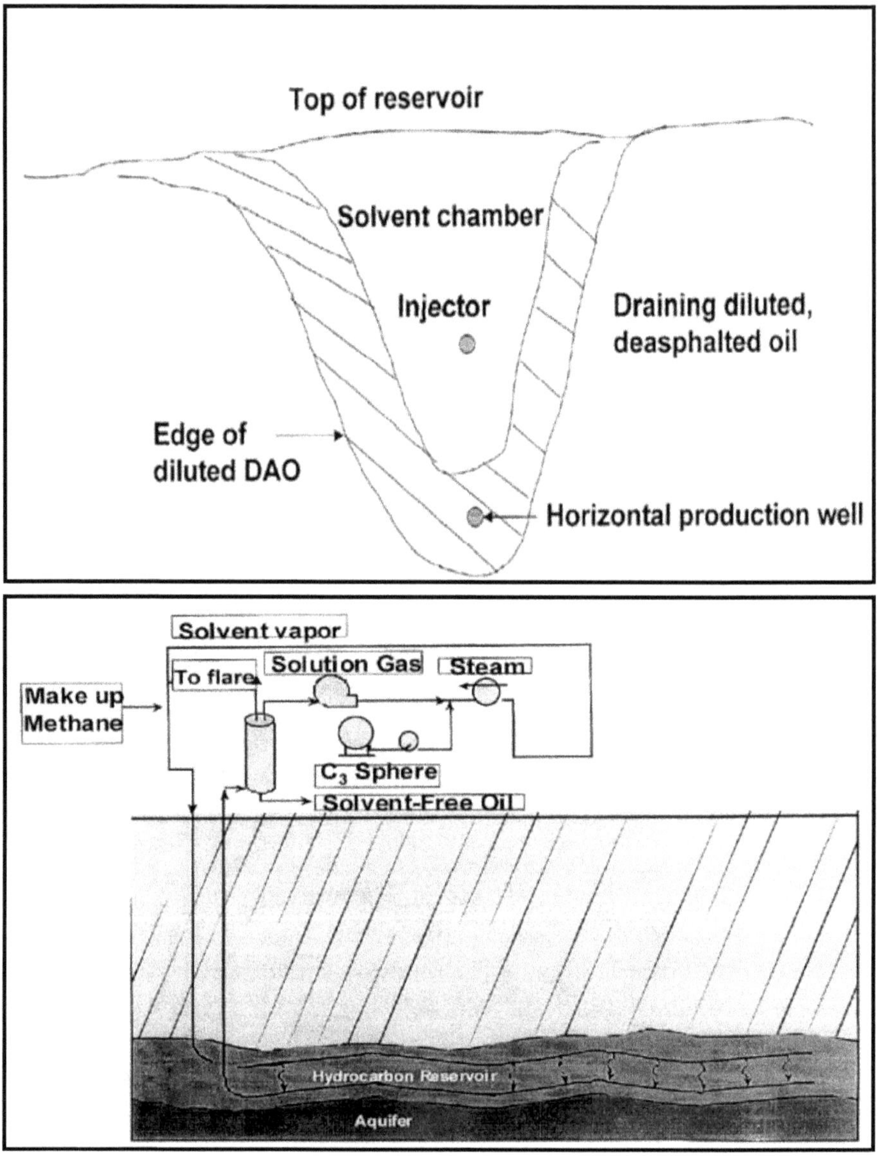

FIGURE 2.13 Well array and surface process diagram for the VAPEX field test (Ovalles 2019).

process reduces GHG emissions during heavy oil recovery because it does not require steam generation, and the reservoir does not need to be heated (Pourabdollah, 2013 and Canbaz 2021). Also, water usage is lower than that of steamflooding.

Lee and Babadagli conducted a comparative analysis of the GHG emissions generated by the solvent extraction process (VAPEX) in comparison with steam-based technologies (Base Case), such as SAGD and steamflooding (Lee 2021).

The results indicated that the direct GHG emissions (fuel use) from VAPEX ranged 5–30 kg CO_2e/BBL, while the emissions for the Base Case (SAGD) varied in the range 60–100 kg CO_2e/BBL (Lee 2021). Thus, a GHG potential emission reduction of medium and a TRL level 6 (field pilot, see **Table 1.2**) can be cautiously assessed for the VAPEX technology, as illustrated in the high-level technology assessment (**Figure 1.4**). More details will be presented in Section 5.1.

2.5.2 Cyclic Propane Injection

Cyclic Propane Injection was field-tested in a reservoir in eastern Venezuela with an API gravity of less than 10° in 2000. This technique involves the downhole injection of a light solvent (propane) under reservoir conditions (130°F and 1000 psi) without using steam, following a conventional "Huff and Puff" process (Ovalles 2019). After the injection phase, the same well operates under production conditions. Once the light solvent is separated, an upgraded hydrocarbon product is obtained.

In the field test, 10,000 barrels of LNG (>80% propane) was injected downhole, and the well was put into production without soaking time. The API gravity and the percentages of the heavy fraction (defined as insolubles in hexane) in the produced oil were monitored, and the results are shown in **Figure 2.14**. Initially, for the first 15 days of the test, the produced oil had an API gravity of 26°–27° (represented by the darker trace) and a minimal amount of heavy fraction (represented by the lighter line). After this initial period, a decrease in API gravity was noted, accompanied by an increase in the heavy fraction of the produced oil. After approximately 50 days, both the API gravity and the heavy fraction returned to their original values of 8.7°API and 9%, respectively. The overall API gravity of the cumulative oil produced during the test was 20°API (Ovalles 2019).

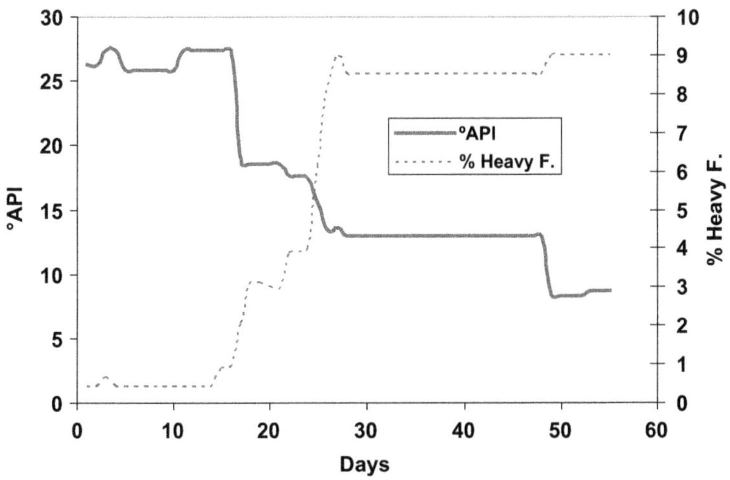

FIGURE 2.14 API gravity and percentage of the heavy fraction of the upgraded crude oil produced during the field test of Cyclic Propane Injection (Ovalles 2019). The percentage of the heavy fraction is defined as insoluble in hexane.

A significant increase in the oil production rate, rising from 50 to 300 BBL/D, was recorded, resulting in a cumulative output of approximately 10,000 BBL of 20° API crude oil over the course of 50 days. From these findings, it can be concluded that the solvent-to-upgraded crude oil ratio is roughly 1:1 (volume/volume) with a 95% solvent recovery rate. Additionally, during the field test, a 97% reduction in water cut and metal removal (vanadium and nickel) was achieved. At the conclusion of the test, no formation damage from asphaltene precipitation was observed, as the original oil production rate of 50 BBL/D was maintained (Ovalles 2019). No new development of this technology has been reported until the publication of this book. Similar to VAPEX, Cyclic Propane Injection eliminates the need for steam to mobilize HCO, significantly reducing water consumption and making it more environmentally friendly than CSS and steamflooding. Currently, only one pilot project has been conducted. Therefore, a TRL levels between 5 and 6 (see **Table 1.2**) can be conservatively assigned, along with a medium potential for reducing GHG emissions, as illustrated in the technology assessment, as shown in **Figure 1.4**. More details will be presented in Section 5.1.

2.5.3 Nsolv™ Technology

This process involves injecting warm light alkanes, such as propane or butane, at approximately 60°C into a SAGD well configuration (see **Figure 2.15**) (Nenninger 2005, 2008, and 2009). When the alkane solvent interacts with the sands underground, where temperatures range between 6°C and 13°C year-round, it condenses, and the bitumen dissolves into the liquefied alkane. By maintaining slightly elevated temperatures, the light alkane warms the bitumen, which accelerates the extraction rate (Nenninger 2005, 2008, and 2009).

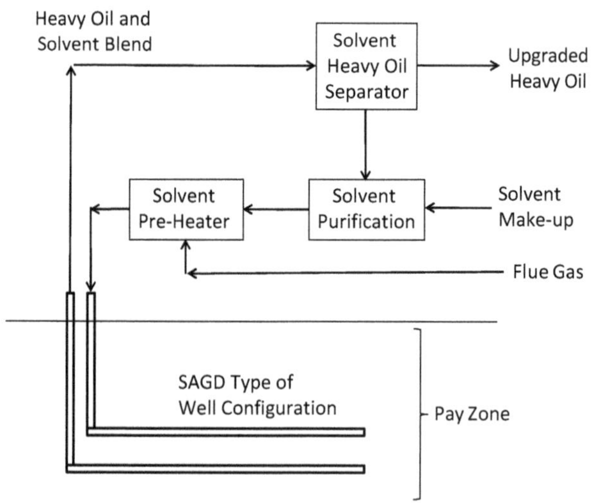

FIGURE 2.15 General process diagram of the Nsolv™ technology (Ovalles 2019).

Due to the high solvent-to-bitumen ratio of 4:1, *in-situ* precipitation of asphaltenes results in the removal of sulfur and metal contaminants. As a result, the produced oil is reported to contain lower levels of sulfur, heavy metals (such as nickel, vanadium, and iron), and carbon residue. This process upgrades the oil's quality to an API gravity of 13°–16°, compared to approximately 8°API for the raw bitumen. Additionally, the produced oil is less viscous, which reduces the amount of diluent needed for pipeline transportation to the refinery (Nenninger 2005, 2008 and 2009).

Nsolv™ initiated field operations with the launch of its pilot plant at Suncor's Dover lease in January 2014. The Bitumen Extraction Solvent Technology (BEST) test facility is designed to process 500 BBL/D. It includes a horizontal well pair that measures 300 m, set in a relatively shallow and thin section of the McMurray formation, which has a 15-m pay zone, 75% oil saturation, 35% porosity, and a permeability of 5 Darcy. The BEST pilot also features a surface plant for processing the extracted hydrocarbons, along with seven observation wells for monitoring and control purposes (Hatch-Nsolv 2025).

In March 2015, the BEST pilot plant started the injection of heated butane into an oil sands reservoir, leading to a partially upgraded oil (from 8° to 9°to 13°–14°API). After two years of operation, the pilot plant produced over 125,000 BBL of upgraded crude oil, while meeting all the following key performance indicators: Extraction rate greater than SAGD produced oil with nickel and vanadium contents less than 125 ppm and Microcarbon residue of ~5wt.%. The solvent recovery is expected to be greater than 90% after blow-down (Krawchuk 2016).

In early 2017, Nsolv™ began the process of shutting down its plant, signaling the conclusion of a highly successful project. The technology developed by Nsolv™ was touted as game-changing, offering significant economic and environmental benefits. The company is actively seeking various partnerships to further develop heavy oil reserves worldwide (Hatch-Nsolv 2025).

K. Wilson from Alberta Innovates published a non-confidential report on the results of the BEST Pilot Project (Wilson 2016). In this report, the author presented a Well-to-Tank LCA demonstrating that a commercial Nsolv™ facility is estimated to reduce *in-situ* extraction emissions by 81% and lower the overall emissions of the fuel supply chain by 32% compared to a SAGD unit of the same capacity. By eliminating water usage, Nsolv™ conserves all the water that would typically be consumed in the SAGD process. The author calculated the GHG emissions intensity for various project scenarios, including the original pre-pilot emissions estimate, emissions based on data from the ramp-up of the pilot facility, and emissions calculated from the commercial-scale model of the Nsolv™ process. The total emissions in each scenario are as follows: 12.7, 38.6, and 16.3 kg CO_2e/BBL for the original, adjusted pilot, and commercial scenarios, respectively. Additionally, the study also examined reductions in emissions of particulate matter, nitrogen oxides, and sulfur oxides (Wilson 2016). Thus, the Nsolv™ process shows high potential for reducing GHG emissions; however, it has only been validated in a single field test, which is classified as TRL level 6 (**Table 1.2**), as illustrated in the technology assessment (see **Figure 1.4**).

2.6 CONCLUSION

- Non-thermal EOR technologies for extracting HCO have been commercially implemented worldwide due to their simplicity and cost-effectiveness, achieving a TRL level 9 with multi-well and multi-field applications. Several emerging, environmentally friendly technological approaches have also been developed (TRL levels between 3 and 9), which reduce GHG emissions while maintaining or improving oil recovery factors by up to 30% of the OOIP.
- Among the advanced processes with high potential for reducing GHG emissions, we can find multilateral wells with or without AICD, EDH, polymer flooding, and subsurface solvent deasphalting. With medium potential, the routes include alkaline-surfactant-polymer flooding, carbon dioxide and nitrogen injection processes, enhanced CHOPS, and solvent-only injection (using both hydrocarbon-based and environmentally friendly solvents).

REFERENCES

Abbaszadeh, M., Ohno, K., Okano, H., 2008, "Reservoir Characterization and CO_2-EOR Injection Pilot Studies in Chicontepec Turbidite Reservoirs, Mexico." Paper presented at the *International Petroleum Technology Conference*, Kuala Lumpur, Malaysia, December. https://doi.org/10.2523/IPTC-12637-MS

Abuov, Y., Serik, G., Lee, W., 2022, "Techno-Economic Assessment and Life Cycle Assessment of CO_2-EOR", *Environ. Sci. Technol.*, 56, 8571–8580, https://doi.org/10.1021/acs.est.1c06834

Acosta, W., Garcés, C.A., 2013, "First Heater Cable Installed in Colombia." Paper presented at the SPE Artificial Lift Conference-Americas, Cartagena, Colombia, May. https://doi.org/10.2118/165064-MS

Advanced Bio-Solvent Utilization for Solvent assisted SAGD Performance Improvement. BioOilSolv, 2025, Emission Reduction Alberta, *BioOilSolv Pilot Test.* Retrieved November 7, 2025.

Ahmadi, M., 2023, *Sustainable In-Situ Heavy Oil and Bitumen Recovery*. Gulf Professional Publishing, Cambridge, MA, USA.

Aitkulov, A., Redwine, C., Alvord, J., Edwards, R., Seright, R.S., 2024, "Polymer Solution Preparation, Quality Control, and Produced Fluid Analysis at Milne Point," *SPE J.*, 29, 6155–6171. https://doi.org/10.2118/218214-PA

Almasri, A., Abdulhalim, A., Mahmoud, A.-G., Farahat, M., 2024, "Enhancing Productivity, Optimizing Well Flow and Reducing Costs of Long Dual Lateral Oil Well Using Integrated Multilateral Drilling and Smart Completions Technologies Coupled with Inflow Control Devices: Best Practices, Lessons Learned, and Value Added." Paper presented at the *GOTECH*, Dubai, UAE, May. https://doi.org/10.2118/219206-MS

Alvarado, V., Manrique, E., 2010, "Enhanced Oil Recovery: An Update Review," *Energies*, 3, 1529–1575; https://doi.org/10.3390/en3091529

Arias, A.S., Navarro, E.G., Munevar, F.C., Ortiz, I.L., 2021, "Sharpening a Cyclic Solvent Injection Technique in a Heavy Oil Reservoir from the Variation of the Nitrogen Injected as Dispersing: A Field Pilot, Colombia." Paper presented at the *SPE Europec featured at 82nd EAGE Conference and Exhibition*, Amsterdam, The Netherlands, October. https://doi.org/10.2118/205228-MS

Arnold, W.A., Graham, R.E., Wagg, B.T., 2012, "Enhanced CHOPS Using SuperSump to Reduce Environmental Footprint and Increase Oil Recovery." Paper presented at the *SPE Heavy Oil Conference Canada*, Calgary, Alberta, Canada, June. https://doi.org/10.2118/157940-MS

Bayestehparvin, B., Farouq Ali, S.M., Abedi,J., 2018, "Solvent-Based and Solvent-Assisted Recovery Processes: State of the Art," SPE No. 179829, *SPE Res. Eval. Eng.*, 22, 29–49, August, 1–21, and references therein. https://doi.org/10.2118/179829-PA

Beckman, M., 2018, "Cyclic Solvent Process Pilot," Non-Confidential Final report for emissions reduction Alberta, Project Completion Date: June 30. https://www.eralberta.ca/wp-content/uploads/2019/01/CSP-Non-Confidential-Report.pdf. Retrieved March 3, 2025.

Berry, S.L., Boles, J.L., Cawiezel, K.E., 2007, "Evaluation of a Renewable, Environmentally Benign Green Solvent for Wellbore and Formation Cleaning Applications." Paper presented at the *International Symposium on Oilfield Chemistry*, Houston, Texas, U.S.A., February. https://doi.org/10.2118/106067-MS

BioOilSolv Technology, 2025, http://biooilsolv.ca/technology/. Retrieved November 7, 2025

Bone, R.E., Aguilar, F.C., Segovia, K.E., 2016, "New Challenges in Construction and Production Prediction Using Multilateral Well to Develop Ecuadorian Oilfield in Environmental Sensitive Areas." Paper presented at the *SPE Trinidad and Tobago Section Energy Resources Conference*, Port of Spain, Trinidad and Tobago, June. https://doi.org/10.2118/180810-MS

British Petroleum Company, 2022, "BP Statistical Review of World Energy," London, Statistical Review of World Energy 2022. Retrieved April 2025.

Butler, R., Yee, C.T., 2000, "Progress in the In Situ Recovery of Heavy Oils and Bitumen." Paper presented at the *Canadian International Petroleum Conference*, Calgary, Alberta, June. https://doi.org/10.2118/2000-050

Bybee, K., 2011, "Alaskan Heavy Oil: First CHOPS at an Untapped Arctic Resource," *J Pet Technol.*, 63, 87–88. https://doi.org/10.2118/0311-0087-JPT

Canbaz, C.H., Temizel, C., Palabiyik, Y., Kor, K., Hendrandingrat, L., Bahar Hosgor, F., Aksahan, F., 2021, "Environmentally-Friendly Production and Recovery Processes for Heavy Presented at *SPE Trinidad and Tobago Section Energy Resources Conference* held virtually 28–30 June 2021. https://doi.org/10.2118/200897-MS

Carbon Capture Magazine, 2023, "Restart of the large-scale Petra Nova Carbon Capture Facility," November 23, Restart of the large-scale Petra Nova Carbon Capture Facility|Carbon Capture Magazine. Accessed April 29, 2025.

Chandler, W.C., 1996, "Industrial solvent based on a processed citrus oil for cleaning up petroleum waste products", US Patent No. 5,549,839, Aug. US5549839A - Industrial solvent based on a processed citrus oil for cleaning up petroleum waste products - Google Patents

Chang, J., Ivory, J., 2015, "History Match and Strategy Evaluation for CSI Pilot." Paper presented at the *SPE Canada Heavy Oil Technical Conference, Calgary*, Alberta, Canada, June. https://doi.org/10.2118/174467-MS

Chang, J., Ivory, J., Forshner, K., Feng, Y., 2013, "Impact of Solvent Loss During Solvent Injection Processes." Paper presented at the *SPE Heavy Oil Conference-Canada, Calgary*, Alberta, Canada, June. https://doi.org/10.2118/165476-MS

Chang, J., Ivory, J., Rajan, R.S.V., 2009, "Cyclic Steam-Solvent Stimulation Using Horizontal Wells", PETSOC-2009-175, presented at the *Canadian International Petroleum Conference*, Calgary, Alberta, 16–18 June. https://doi.org/10.2118/2009-175

Chung, F.T.H., 1988, "Oil recovery by Immiscible CO_2 Flooding-Topical Report" IIT Research Institute, *Nat. Inst. Pet. & Ener. Res. NIPER 297*, Bartlesville. OK March. CO_2 Flooding-Topical Report

Clean Air Task Force, 2019, "CO_2 EOR Yields a 37% Reduction in CO_2 Emitted Per Barrel of Oil Produced," June 7, 2019, Clean Air Task Force. Retrieved November. 4, 2025.

CORDIS - EU research results, 1985, "Pilot Field for Enhancement of Heavy Oil Recovery Ponte Dirillo Field-Italy," Grant agreement ID: TH.-05014-79, May, 29. https://cordis.europa.eu/project/id/TH.-05014-79. Retrieved July 2, 2025.

Curtis, J., 2003, "Environmentally Favorable Terpene Solvents Find Diverse Applications in Stimulation, Sand Control and Cementing Operations." Paper presented at the *SPE Annual Technical Conference and Exhibition*, Denver, Colorado, October. https://doi.org/10.2118/84124-MS

Das, S.K., 1998, "Vapex: An Efficient Process for the Recovery of Heavy Oil and Bitumen," *SPE J.*, 3 232–237. https://doi.org/10.2118/50941-PA

Das, S., 2005, "Diffusion and Dispersion in the Simulation of VAPEX Process", SPE No. 97924, presented at *SPE International Thermal Operation and Heavy Oil Symposium*, Calgary Alberta, November. 1–3 and references therein. https://doi.org/10.2118/97924-MS

de Boer, R.B., Leerlooyer, K., Eigner, M.R.P., van Bergen A.R.D., 1995, "Screening of Crude Oils for Asphalt Precipitation: Theory, Practice, and the Selection of Inhibitors," *SPE Prod & Fac.*, 10, 55–61. https://doi.org/10.2118/24987-PA

Delamaide, E., 2014, "Polymer Flooding of Heavy Oil - From Screening to Full-Field Extension." Paper presented at the *SPE Heavy and Extra Heavy Oil Conference: Latin America*, Medellín, Colombia, September. https://doi.org/10.2118/171105-MS

Delamaide, E., 2025, "Current Status of Polymer Flooding in Heavy Oil Fields: When Performances Beat Theory." Paper presented at the *SPE Western Regional Meeting*, Garden Grove, California, USA, April. https://doi.org/10.2118/224158-MS

Delamaide, E., Bazin, B., Rousseau, D., Degre, G., 2014b, "Chemical EOR for Heavy Oil: The Canadian Experience." Paper presented at the *SPE EOR Conference at Oil and Gas West Asia*, Muscat, Oman, March 2014. https://doi.org/10.2118/169715-MS

Delamaide, E., Zaitoun, A., Renard, G., Tabary, R., 2014a, "Pelican Lake Field: First Successful Application of Polymer Flooding In a Heavy-Oil Reservoir," *SPE Res Eval & Eng.*, 17, 340–354. https://doi.org/10.2118/165234-PA

Department of Energy, 2020, "Happy Third Operating Anniversary, Petra Nova!", *Office of Fossil Energy and Carbon Management*, January 10. Happy Third Operating Anniversary, Petra Nova!|Department of Energy. Accessed April 29, 2025.

Du, Z., Zeng, F., Chan, C., 2013, "An Experimental Study of the Post-CHOPS Cyclic Solvent Injection Process." Paper presented at the *SPE Heavy Oil Conference-Canada, Calgary*, Alberta, Canada, June https://doi.org/10.2118/165524-MS

Dubin, K., 2017, "Petra Nova is One of Two Carbon Capture and Sequestration Power Plants in the World", October 31, Petra Nova is one of two carbon capture and sequestration power plants in the world - U.S. Energy Information Administration (EIA). Accessed April 29, 2025

Dupuis, G., Niewerf, J., 2020, "A Cost-Effective EOR Technique To Reduce Carbon Intensity With Polymer Flooding and Modular Skids," *J. Pet. Tech.*, August 25, https://jpt.spe.org/cost-effective-eor-technique-reduce-carbon-intensity-polymer-flooding-and-modular-skids

Dusseault, M.B., Liang, C.X., Ma, Y., Wu, G., Xu, B., 2002, "CHOPS in Jilin Province, China." Paper presented at the *SPE International Thermal Operations and Heavy Oil Symposium and International Horizontal Well Technology Conference*, Calgary, Alberta, Canada, November. https://doi.org/10.2118/79032-MS

Dutta, S.J., Dhrubajyoti, N., 2025, "Sustaining Heavy Crude Oil Production with Chemical EOR: A Comparative Study," *Chem. Therm. Thermal Anal.*, 18, 100191–100205. https://doi.org/10.1016/j.ctta.2025.100191

Elturki, M., Abdulmohsin A., 2022, "Asphaltene Precipitation and Deposition under Miscible and Immiscible Carbon Dioxide Gas Injection in Nanoshale Pore Structure," *SPE J.*, 27, 3643–3659. https://doi.org/10.2118/210592-PA

Espinosa, M.G., Zavala-Arriaga, M., Ramírez-Gonzalez, P.V., 2022, "Enhanced Oil Recovery with D-Limonene Diluted in Brine," *J. Pet. Sci. Eng.*, 210, March, 110110–110121. https://doi.org/10.1016/j.petrol.2022.110110

Fassihi, M.R., Kovscek, A.R., 2017, "Low-Energy Processes for Unconventional Oil Recovery" *Society of Petroleum Engineers*, Richardson, Texas, USA, Section 5.2. ISBN 978-1-61399-475-7. https://doi.org/10.2118/9781613994757

Fath, A.H., Pouranfard, A.-R., 2014, "Evaluation of Miscible and Immiscible CO_2 Injection in One of the Iranian Oil Fields," *Egyptian J. Pet.*, 23, 255–270. https://doi.org/10.1016/j.ejpe.2014.08.002

Firouz, A.Q., Torabi, F., 2012, "Feasibility Study of Solvent-Based Huff-n-Puff Method (Cyclic Solvent Injection) to Enhance Heavy Oil Recovery." Paper presented at the *SPE Heavy Oil Conference Canada*, Calgary, Alberta, Canada, June. https://doi.org/10.2118/157853-MS

Fischer, K. 2013, "Terpenes Replacing BTEX In Oil Field", The American Oil & Gas Reporter, Sept. Terpenes Replacing BTEX In Oil Field. Retrieved July 21, 2025

Gandomkar, A., Torabi, F., Nasriani, H.R., Enick, R.M., 2023, "Decreasing Asphaltene Precipitation and Deposition During Immiscible Gas Injection Via the Introduction of a CO_2-Soluble Asphaltene Inhibitor," *SPE J.*, 28, 2316–2328. https://doi.org/10.2118/214698-PA

Ganguli, S.S., 2017, "Integrated Reservoir Studies for CO_2-Enhanced Oil Recovery and Sequestration Application to an Indian Mature Oil Field," *Springer*, https://doi.org/10.1007/978-3-319-55843-1

Garnett, R.L., 2001, "Recovery of Heavy Oil from the Monterey Formation in Offshore California by Cyclic Injection of Light-Oil Diluent," *SPE Re.s Eval. & Eng.*, 4, 51–58. https://doi.org/10.2118/70992-PA

Garnier, O., Blanchet, M., D'Alessandro, S., Bourgeois, M., Prinet, C., Romero, C., Vaswani, Y., Aljaberi, M., Elbekshi, A., Al Saadi, H.A., Hegazy, O., Al Daghar, K., Alhammadi, A., 2023, "A Novel Workflow Proposed to Calculate Net Carbon Intensity for EOR Applications in Giant Middle East Oil Reservoir." Paper presented at the *ADIPEC*, Abu Dhabi, UAE, October. https://doi.org/10.2118/216643-MS

Goavec, M., Salaün, M., Morvan, M., Delamaide, E., Rousseau, D., Negre A., 2023, "Towards Sustainable Oil Production: CO_2 Footprint Assessment of EOR Surfactant-Based Processes." Paper presented at the *SPE International Conference on Oilfield Chemistry*, The Woodlands, Texas, USA, June 2023. https://doi.org/10.2118/213813-MS

Gomaa, S., Salem, K.G., El-Hoshoudy, A.N., 2024, "Enhanced Heavy and Extra Heavy Oil Recovery: Current Status and New Trends," *Petroleum*, 10(3), 399–410. https://doi.org/10.1016/j.petlm.2023.10.001

Gondiken, S., 1987, "Camurlu Field Immiscibie CO_2 Huff and Puff Pilot Project" Paper presented at the *Middle East Oil Show*, Bahrain, March. https://doi.org/10.2118/15749-MS

Guo, K., Li, H., Yu, Z., 2016, "In-Situ Heavy and Extra-Heavy Oil Recovery: A Review," *Fuel*, 185, 886–902, and references therein. https://doi.org/10.1016/j.fuel.2016.08.047

Gutierrez, M., Castro, R.H., Corredor, L.M., Fernández, F.R., Zapata, J., Jimenez, J.A., Reyes, J.D., Rojas, D.M., Jimenez, R, Acosta, T., Dueñas, D.E., Solorzano, P.L., Mayorga, H., Llanos, S., Quintero, H.I., Garcia. H.A., 2024, "Chemical Enhanced Oil Recovery Experiences in Colombia: Field Pilots Review." Paper presented at the *SPE Improved Oil Recovery Conference*, Tulsa, Oklahoma, USA, April. https://doi.org/10.2118/218173-MS

Haddad, A.S., Gates, I., 2017, "CO_2-Based Heavy Oil Recovery Processes for Post-CHOPS Reservoirs," *J. CO_2 Utiliz.*, 19, 238–246, https://doi.org/10.1016/j.jcou.2017.03.019

Hammami, A., Phelps, C.H., Monger-McClure, T., Little, T. M., 1999, "Asphaltene Precipitation from Live Oils: An Experimental Investigation of Onset Conditions and Reversibility," *Energy Fuels*, 14(1), 14–18. https://doi.org/10.1021/ef990104z

Hatch-Nsolv Website, 2025, https://www.hatch.com/Projects/Energy/Nsolv-Pilot-Plant. Accessed on May 12.

Huerta, M., Otero, C., Rico, A., Jimenez, I., de Mirabal, M., Rojas, G., 1996, "Understanding Foamy Oil Mechanisms for Heavy Oil Reservoirs During Primary Production." Paper presented at the *SPE Annual Technical Conference and Exhibition*, Denver, Colorado, October. https://doi.org/10.2118/36749-MS

Islam M.R., 2020, *Economically and Environmentally Sustainable Enhanced Oil Recovery*. Wiley, Hoboken, NJ, USA. ISBN 978-1-119-47909-3

Islam, M.R., Chakma, A., 1990, "Mechanics of Bubble Flow in Heavy Oil Reservoirs" SPE No. 20070 presented at the *SPE California Regional Meeting*, Venture, CA, 4–6 April. https://doi.org/10.2118/20070-MS

Ivory, J., Chang, J., Coates, R., Forshner, K, 2010, "Investigation of Cyclic Solvent Injection Process for Heavy Oil Recovery," *J Can Pet Technol*, 49, 2–33. https://doi.org/10.2118/140662-PA

Jamaloei, B.Y., Dong, M., Mahinpey, N., Maini, B.B., 2012, "Enhanced Cyclic Solvent Process (ECSP) for Heavy Oil and Bitumen Recovery in Thin Reservoir," *Energy & Fuels*, 26(5), 2865–2874. https://doi.org/10.1021/ef300152b

Jamaluddin, A.K.M., Joshi, N., Iwere, F., Gurpinar, O, 2022, "An Investigation of Asphaltene Instability Under Nitrogen Injection." Paper presented at the *SPE International Petroleum Conference and Exhibition in Mexico*, Villahermosa, Mexico, February. https://doi.org/10.2118/74393-MS

Janssen, M.T., Azimi, F., Zitha, P.L., 2018, "Immiscible Nitrogen Flooding in Bentheimer Sandstones: Comparing Gas Injection Schemes for Enhanced Oil Recovery." Paper presented at the *SPE Improved Oil Recovery Conference*, Tulsa, Oklahoma, USA, April. https://doi.org/10.2118/190285-MS

Jia, X., Zeng, F., Gu, Y., 2013, "Pressure Pulsing Cyclic Solvent Injection (PP-CSI): A New Way to Enhance the Recovery of Heavy Oil through Solvent-Based Enhanced Oil Recovery Techniques." Paper presented at the *SPE Annual Technical Conference and Exhibition*, New Orleans, Louisiana, USA, September. https://doi.org/10.2118/166453-MS

Jiang, T., Jia, X., Zeng, F., Gu, Y., 2013 "A Novel Solvent Injection Technique for Enhanced Heavy Oil Recovery: Cyclic Production with Continuous Solvent Injection." Paper presented at the *SPE Heavy Oil Conference-Canada*, Calgary, Alberta, Canada, June 2013. https://doi.org/10.2118/165455-MS

JIVE: Joint Implementation of Vapor Extraction Project, 2010, Petroleum Technology Research Centre (PTRC), https://ptrc.ca/past-projects/jive. Retrieved July 17, 2025.

Joshi, N.B., Mullins, O.C., Jamaluddin, A. Creek, J., McFadden, J., 2001 "Asphaltene Precipitation from Live Crude Oil," *Energy Fuels*, 15(4), 979–986. https://doi.org/10.1021/ef010047l

Konopczynski, M., Grefa, M., Castillo, M., Villavicencio, A., Solano, M., Chanaga, A., 2024 "Improving Production and Hydrocarbon Recovery from Viscous Oil Reservoirs in Ecuador While Minimizing the Carbon Intensity of Production Using AICD Technology." Paper presented at the *SPE Western Regional Meeting*, Palo Alto, California, USA, April https://doi.org/10.2118/218869-MS

Krawchuk P., 2016, "Development and Operation of the Nsolv BEST Demonstration Facility". Paper No. WHOC16-135, presented at the 2016 World Heavy Oil Congress, Calgary, Alberta, Canada, September 6–9.

Kumar, R., Kakati, P., Lone, A.S., Bharadwaj, S., Gurjar, S.S., Pathak, D., 2023, "Case Study for Enhancement of Production of Heavy and Highly Viscous Crude Oil Using Electrical Downhole Heater." Paper presented at the *SPE/IATMI Asia Pacific Oil & Gas Conference and Exhibition*, Jakarta, Indonesia, October. https://doi.org/10.2118/215309-MS

Lechtenberg, H.L., Gates, W.L., Caraway, H., Baptist, O.C., 1974, "Solvent Stimulation Tests in Two California Oilfields," Washington, US Bureau of Mines, TN23.U7. https://babel.hathitrust.org/cgi/pt?id=mdp.39015078484345&seq=3. Retrieved Jul. 17, 2025

Lee, J., Babadagli, T., 2021, "Mitigating Greenhouse Gas Intensity Through New Generation Techniques During Heavy Oil Recovery," *J.Clean Prod.*, 286, 124980. https://doi.org/10.1016/j.jclepro.2020.124980

Leontaritis, K.J., Mansoori, G.A., 1988, "Asphaltene Deposition: A Survey of Field Experiences and Research Approaches," *J. Pet. Sci. Eng.*, 1(3), 229–239. https://doi.org/10.1016/0920-4105(88)90013-7

Li, K., Ovsepian, M., Xie, W., Varfolomeev, M.A., Luo, Q., Yuan, C., 2024, "Emulsions for Enhanced Oil Recovery: Progress and Prospect," *J. Mol. Liquid.*, 393(1), 12358–12367. https://doi.org/10.1016/j.molliq.2023.123658

Lim, G.B., Kry, R.P., Harker, B.C., Jha, K.N., 1995, "Cyclic Stimulation of Cold Lake Oil Sand with Supercritical Ethane." Paper presented at the *SPE International Heavy Oil Symposium*, Calgary, Alberta, Canada, June. https://doi.org/10.2118/30298-MS

Lin, L., Ma, H., Zeng, F., Gu, Y., 2014, "A Critical Review of the Solvent-Based Heavy Oil Recovery Methods." Paper presented at the *SPE Heavy Oil Conference-Canada*, Calgary, Alberta, Canada, June 2014. https://doi.org/10.2118/170098-MS

Mahdavi, E., Zebarjad, F.S., 2018, "Screening Criteria of Enhanced Oil Recovery Methods. Fundamentals of Enhanced Oil and Gas Recovery from Conventional and Unconventional Reservoirs," 41–59. https://doi.org/10.1016/b978-0-12-813027-8.00002-3

Mai, A., Bryan, J., Goodarzi, N., Kantzas, A., 2009, "Insights into Non-Thermal Recovery of Heavy Oil," *J Can Pet Tech.*, 48(03), 27–35. Paper Number: PETSOC-09-03-27 https://doi.org/10.2118/09-03-27

Masnadi, M.S., El-Houjeiri, H.M., Schunack, D., Li, Y., Englander, J.G., Badahdah, A., Monfort, J.-C., Anderson, J.E., Wallington, T.J., Bergerson, J.A., Gordon, D., Koomey, J., Przesmitzki, S., Azevedo, I.L., Bi, X.T., Duffy, J.E., Heath, G.A., Keoleian, G.A., McGlade, C., Meehan, D.N., Yeh, S., You, F., Wang, M., Brandt, A.R., 2018, "Global Carbon Intensity of Crude Oil Production," *Science*, 361(6405), 851–853. https://doi.org/10.1126/science.aar6859

Mathews, T.A., Azzu, P., Cortes, J., Hascakir, B., 2022, "Effective Extraction of a Heavy Oil Resource by an Environmentally Friendly Green Solvent: Limonene." Paper presented at the *SPE Annual Technical Conference and Exhibition*, Houston, Texas, USA, October. https://doi.org/10.2118/210138-MS

Mathews, T.A., Azzu, P., Hascakir, B., 2024, "Unlocking Sustainable Solutions: Exploring Terpene-Based Green Solvents for Enhanced Heavy Oil Recovery," *Energy & Fuel*, 38, 23336–23347, https://doi.org/10.1021/acs.energyfuels.4c02793

McAuliffe, C.D., 1973, "Crude-Oil-Water Emulsions to Improve Fluid Flow in an Oil Reservoir," *J Pet Technol*, 25, 721–726. https://doi.org/10.2118/4370-PA

Medizade, M.M., Wiltshire, Z., 2019, "Further Investigation of Downhole Heating by an Electric Cable." Paper presented at the *SPE Western Regional Meeting*, San Jose, California, USA, April. https://doi.org/10.2118/195324-MS

Mirabal, M., Gordillo, R., Rojas, G., Rodriguez, H., Huerta, M., 1996, "Impact of Foamy Oil Mechanism on the Hamaca Oil Reserves, Orinoco Belt-Venezuela." Paper presented at the *SPE Latin America/Caribbean Petroleum Engineering Conference*, Port-of-Spain, Trinidad, April. https://doi.org/10.2118/36140-MS

Moradi, M., García, W., Amado, P.M., Konopczynski, M.R., 2023, "Delivering NET ZERO–A Case Study of Minimized Carbon Intensity Production Using Autonomous Inflow Control Technologies from a Remote Location in the Peruvian Amazon." Paper presented at the *SPE Europe- Europe Energy Conference featured at the 84th EAGE Annual Conference & Exhibition*, Vienna, Austria, June. https://doi.org/10.2118/214343-MS

Morales, R., Rodriguez, F., Bejarano, C., Belhaj, H., Alhameli, F., AlDhuhoori, M., Quintero, E., 2024, "IOR/EOR Project Implementation Outcomes and Novel Insights in Extra-Heavy Oil Reservoirs in Eastern Venezuela Basin, Orinoco Oil Belt – Zuata Principal Field, Venezuela" Paper presented at the *ADIPEC*, Abu Dhabi, UAE, November. https://doi.org/10.2118/222810-MS

Morgan, A., Ampomah, W., Grigg, R., Dai, Z., You, J., Wang, S., 2023, "Techno-Economic Life Cycle Assessment of CO_2-EOR Operations Towards Net Negative Emissions at Farnsworth Field Unit," *Fuel*, 342 12897–12909, https://doi.org/10.1016/j.fuel.2023.127897

Naderi, K., Babadagli, T., 2007, "Clarifications on Oil/Heavy Oil Recovery Under Ultrasonic Radiation Through Core and 2D Visualization Experiments". Presented at the *Petroleum Society's 8th Canadian International Petroleum Conference* (58th Annual Technical Meeting), Calgary, Alberta, Canada, June 12–14. https://doi.org/10.2118/2007-032

Nagy, R., Hartyányi, M., Bejczi, R. 2025 "Recent Aspects of Chemical Enhanced Oil Recovery," *Chem. Pap.*, https://doi.org/10.1007/s11696-025-03943-7

Nenninger, J.E., Dunn, S.G. 2008 "How Fast is Solvent Based Gravity Drainage?." Paper presented at the *Canadian International Petroleum Conference*, Calgary, Alberta, June. https://doi.org/10.2118/2008-139

Nenninger, J.E., Gunnewiek, L., 2009, "Dew Point vs Bubble Point: A Misunderstood Constraint on Gravity Drainage Processes." Paper presented at the *Canadian International Petroleum Conference*, Calgary, Alberta, June. https://doi.org/10.2118/2009-065

Nenninger, J., Nenninger, E., 2005, "Method and Apparatus for Stimulating Heavy Oil Production", US Patent 6,883,607. https://patentimages.storage.googleapis.com/0e/ee/c8/9a4e0607e4993f/US6883607.pdf

Nwosi-Anele, A.S., Obuebite, A.A., Okoro, E.E., Onyekonwu, M., 2019, "Heavy Oil Recovery: Environmental Implications and Mitigation Measures," *J. Oil, Gas and Coal Eng.*, 4(1), 34–38. https://doi.org/10.2118/172400-ms

Ojeda, S.S., Parman, D.G., 2013, "Use of Electric Downhole Heaters to Improve Production and Recovery of Heavy, Viscous Oil in California and Venezuela." Paper presented at the *SPE Kuwait Oil and Gas Show and Conference*, Kuwait City, Kuwait, October. https://doi.org/10.2118/167347-MS

Olajire, A. A., 2014, "Review of ASP EOR (alkaline surfactant polymer enhanced oil, recovery) Technology in the Petroleum Industry: Prospects and Challenges," *Energy*, 77, 963–982. https://doi.org/10.1016/j.energy.2014.09.005

Orr, F.M., 2007, "Theory of Gas Injection," Tie-line Publication, Copenhagen, Denmark. ISBN: 9788798996125. https://tie-tech.com/vare/theory-of-gas-injection-processes/

Ovalles, C., 2019, *Subsurface Upgrading of Heavy Crude Oils and Bitumen*. C. Ovalles, CRC Press, Boca Raton, Florida, USA, August 8th. https://doi.org/10.1201/9781315181028

Ovalles, C., Fonseca, A., Lara, A., Alvarado, V., Urrecheaga, K., Ranson, A., Mendoza, H., 2002, "Opportunities of Downhole Dielectric Heating in Venezuela: Three Case Studies Involving Medium, Heavy and Extra-Heavy Crude Oil Reservoirs." Paper presented at the *SPE International Thermal Operations and Heavy Oil Symposium and International Horizontal Well Technology Conference*, Calgary, Alberta, Canada, November. https://doi.org/10.2118/78980-MS

Ovalles, C., Rogel, E., Alboudwarej, H., Inouye, A., Benson, I. P., Vaca, P., 2017, "Physical and Numerical Simulations of Subsurface Upgrading by Use of Solvent Deasphalting in a Heavy-Crude-Oil Reservoir," *SPE Res Eval & Eng*, 20, 654–668. https://doi.org/10.2118/183636-PA

Ovalles, C., Rogel, E., Moir, M.E., Thomas, L., Pradhan, A., 2012, "Characterization of Heavy Crude Oils, Their Fractions, and Hydrovisbroken Products by the Asphaltene Solubility Fraction Method," *Energy & Fuel*, 26, 549. https://doi.org/10.1021/ef201499f

Panggabean, Y.B., Pramadewi, T.B., Kusuma, E.A., Maryanto, A., 2022, "Designing CO_2-EOR in Indonesia by Matching Business Strategy: Study Case East Java Field, Indonesia." Paper presented at the *SPE EuropEC - Europe Energy Conference featured at the 83rd EAGE Annual Conference & Exhibition*, Madrid, Spain, June. https://doi.org/10.2118/209704-MS

Pinilla, A., Asuaje, M., Ratkovich, N., 2025, "Comparison Between AICV, ICD, and Liner Completions in the Displacement Front and Production Efficiency in Heavy Oil Horizontal Wells," *Processes*, 13, 1576. https://doi.org/10.3390/pr13051576

Pizarro, J.O.S., Trevisan. O. V., 1990, "Electrical Heating of Oil Reservoirs: Numerical Simulation and Field Test Results," *J. Pet. Tech.*, 42, 1320–1326. https://doi.org/10.2118/19685-PA

Pooladi-Darvish, M., Firoozabadi, A., 1999, "Solution-gas Drive In Heavy Oil Reservoirs," *J. Can. Pet. Techno*, 38, no pagination specified, https://doi.org/10.2118/99-04-06

Pourabdollah, K., Mokhtari, B., 2013, "The VAPEX Process, from Beginning Up to Date," *Fuel*, 107, 1–33, and references therein. https://doi.org/10.1016/j.fuel.2012.12.003

Quintero, E., Quezada, A., Romero, I., 2014, "Technical-Economic Evaluation of Downhole Electric Heater Cable Application in Orinoco Oil Belt: Successful Case." Paper presented at the *SPE Latin America and Caribbean Petroleum Engineering Conference*, Maracaibo, Venezuela, May. https://doi.org/10.2118/169351-MS

Quintero, H.I., Solorzano, P.L., Barbosa, C., Corredor, L.M., Martinez, A., Feriz, E.F., Hernandez, R., Vega, S.M., Carrasca L.K.J., Zapata, J.F., Guerrero, C.M., Dueñas, D., Rojas, C., 2024, "Carbon Footprint and Energy Intensity Assessment for Enhanced Oil Recovery (EOR) Based on Polymer Injection: A Colombian Case Study." Paper presented at the *SPE Improved Oil Recovery Conference*, Tulsa, Oklahoma, USA, April. https://doi.org/10.2118/218169-MS

Ramadhan, R., Mon, M.T., Tangparitkul, S., Tansuchat, R., Agustin, D.A., 2024, "Carbon Capture, Utilization, and Storage in Indonesia: An Update on Storage Capacity, Current Status, Economic Viability, and Policy," *Energy Geoscience*, 5, (4), 100335, https://doi.org/10.1016/j.engeos.2024.100335

Rodriguez, F.M., 2021, "Review of chemical EOR projects in Venezuela: From light to extra-heavy oil reservoirs", *40th Int. Conf. on Ocean, Offshore and Arctic Eng., OMAE 2021*, 10, 21 June–30 June, https://doi.org/10.1115/OMAE2021-63529

Rodriguez, F., Belhaj, H., Morales, R., Velasquez, R., AlDhuhoori, M., Alhameli, F., 2024, "Electrical/Electromagnetic Enhanced Oil Recovery for Unconventional Reservoirs: A Review of Research and Field Applications in Latin America." Paper presented at the *SPE Conference at Oman Petroleum & Energy Show*, Muscat, Oman, April. https://doi.org/10.2118/218665-MS

Rodriguez, F., Delamaide, E., Rousseau, D., Bekri, S., 2022, "Which is the Most Attractive IOR Method to Produce the Venezuelan Highly Viscous Oil Resources in the Energy Transition Era? A Comprehensive Review of Research and Field Applications." Paper presented at the *ADIPEC*, Abu Dhabi, UAE, October. https://doi.org/10.2118/211344-MS

Ruifeng, W., Xintao, Y., Xueqing, T., Xianghong, W., Xinzheng, Z., Li, W., Xiaoling, Y., 2011, "Successful Cold Heavy Oil Production with Sand (CHOPS) Application in Massive Heavy Oil Reservoir in Sudan: A Case Study." Paper presented at the *SPE Heavy Oil Conference and Exhibition*, Kuwait City, Kuwait, December. https://doi.org/10.2118/150540-MS

Saboorian-Jooybari, H., Morteza, D., Chen, Z., 2016, "Heavy Oil Polymer Flooding from Laboratory Core Floods to Pilot Tests and Field Applications: Half-Century Studies," *J. Pet. Sci, Eng.*, 142, 85–100. https://doi.org/10.1016/j.petrol.2016.01.023

Sahin, S., Kalfa, U., Celebioglu, D., 2012, "Unique CO_2-Injection Experience in the Bati Raman Field May Lead to a Proposal of EOR/Sequestration CO_2 Network in the Middle East," *SPE Econ & Mgmt*, 4, 42–50. https://doi.org/10.2118/139616-PA

Sanyal, T., Al-Sammak, I. 2011 "Analysis of the First CHOPS Pilot for Heavy Oil Production in Kuwait." Paper presented at the *Canadian Unconventional Resources Conference*, Calgary, Alberta, Canada, November https://doi.org/10.2118/148966-MS

Seidy-Esfahlan, M., Tabatabaei-Nezhad, S.A., Khodapanah, E., 2024, "Comprehensive Review of Enhanced Oil Recovery Strategies for Heavy Oil and Bitumen Reservoirs in Various Countries: Global Perspectives Challenges, and Solutions," *Heliyon*, 10, e37826. https://doi.org/10.1016/j.heliyon.2024.e37826

Shafiai, S.H., Adel Gohari, A., 2020, "Conventional and Electrical EOR Review: The Development Trend of Ultrasonic Application in EOR," *J. Pet. Expl. Prod. Tech.*, 10, 2923–2945, https://doi.org/10.1007/s13202-020-00929-x

Shen, C., Batycky, J.P., 1999, "Observation of Mobility Enhancement of Heavy Oils Flowing Through Sand Pack Under Solution Gas Drive," *J. Can. Pet. Tech.*, 38(4), 46–53. https://doi.org/10.2118/99-04-05

Sheng, J. J., 2013, *ASP Fundamentals and Field Cases Outside China, Enhanced Oil Recovery Field Case Studies*. Gulf Professional Publishing, Elsevier, pp. 189–201. https://doi.org/10.1016/B978-0-12-386545-8.00009-9

Skauge, T., Skauge, A., Lugo, N., Johnson, G., 2024, "How Polymer Flooding Reduces CO_2 Emissions and Energy Consumption - An Exergy Return on Exergy Investment Case Study." Paper presented at the *SPE Improved Oil Recovery Conference*, Tulsa, Oklahoma, USA, April. https://doi.org/10.2118/218231-MS

Society of Petroleum Engineers (SPE), 2025a, "Combining CHOPS and Other Production Technologies," January 27, https://doi.org/10.2118/PW0147. Accessed April 21, 2025.

Society of Petroleum Engineers (SPE), 2025b, "CHOPS Sand Management," January 27, https://doi.org/10.2118/PW0127. Accessed April 21, 2025.

Society of Petroleum Engineers (SPE), 2025c, "Miscible CO_2 Flooding," February 5, https://doi.org/10.2118/PW0593. Accessed April 23, 2025.

Sofla, M.J.D., Farahani, Z.D., Ghorbanizadeh, S., Namdar, H., 2024, "Experimental Study of Asphaltene Deposition During CO_2 and Flue Gas Injection EOR Methods Employing a Long Core," *Sci. Rep.*, 14, 3772–3783. https://doi.org/10.1038/s41598-024-54395-0

Soiket, Md.I.H., Oni, A.O., Kumar, A., 2019, "The Development of a Process Simulation Model for Energy Consumption and Greenhouse Gas Emissions of a Vapor Solvent-Based Oil Sands Extraction and Recovery Process," *Energy*, 173, 799–808, https://doi.org/10.1016/j.energy.2019.02.109

Sonoro Signal, 2024, "Taking CHOPS Horizontal in Multi-Lined, Multilateral Wells (Part 1)," June 4, Sonoro Signal: Taking CHOPS horizontal in multi-lined, multilateral wells (Part 1) — Sonoro Energy. Accessed April 4, 2025.

Speight, J.G., 2016, "Introduction to Enhanced Recovery Methods for Heavy Oil and Tar Sands" Chapter 8: Nonthermal Methods of Recovery. Gulf Professional Publishing, pp 353–403. ISBN:9780128017654

Sun, X., Zhao, H., Zhang, Y., Liu, Y., Chen, G., Wang, W., 2019, "An Experimental Study on the Oil-Soluble Surfactant-Assisted Cyclic Mixed Solvent Injection Process for Heavy Oil Recovery After Primary Production," *Fuel*, 254, 115656. https://doi.org/10.1016/j.fuel.2019.115656

ThermoFischer Scientific, 2025, "SAFETY DATA SHEET". Poly(acrylamide), granular (non-ionic), approx. M.W. 5 to 6.000.000, December, 19. https://www.fishersci.com/store/msds?partNumber=AC178040050&productDescription=POLY%28ACRYLAMIDE%29+GRANULAR+5G&vendorId=VN00032119&countryCode=US&language=en&form=MG0AV3. Accessed March 19, 2026.

Tileuberdi, N., Mashrapova, M., Toktarbay, Z., 2023, "A Review on Nitrogen Flooding for Enhanced Oil Recovery," *ES Mater. Manuf.*, 22, 968–980. https://doi.org/10.30919/esmm968

Ucan, S., Savoy, E., Federici, M.A., Azcurra, M.O., 2014, "Electrical Wellbore Heating in Environmentally Sensitive Llancanelo Field, Argentina." Paper presented at the *SPE Latin America and Caribbean Petroleum Engineering Conference*, Maracaibo, Venezuela, May. https://doi.org/10.2118/169326-MS

Velasquez, R., Lara, V., Korabelnikov, A., Manapov, T., 2020, "Downhole Electrical Heating Dynamic Model Assessment in Faja Extraheavy Oil Reservoir Application." Paper presented at the *SPE Latin American and Caribbean Petroleum Engineering Conference*, Virtual, July. https://doi.org/10.2118/199166-MS

Villarroel, T., Hernández, R., 2013,"Technological Developments for Enhancing Extra Heavy Oil Productivity in Fields of the Faja Petrolifera del Orinoco (FPO), Venezuela", presented at *AAPG Annual Convention and Exhibition*, Pittsburgh, Pennsylvania, May 19–22, Technological Developments FPO. Accessed April, 23, 2025.

Wang, X., Zou, H., Li, G., Nie, C., Chen, J., 2005, "Integrated Well-Completion Strategies With CHOPS To Enhance Heavy-Oil Production: A Case Study in Fula Oilfield." Paper presented at the *SPE International Thermal Operations and Heavy Oil Symposium*, Calgary, Alberta, Canada, November. https://doi.org/10.2118/97885-MS

Wei, Y., Lu, X., Xu, J., 2024, "A Systematical Review of the Largest Alkali-Surfactant-Polymer Flood Project in the World: From Laboratory to Pilots and Field Application," *SPE J.*, 29, 4147–4165. https://doi.org/10.2118/215058-PA

Wilson, A., 2015, "Pelican Lake: First Successful Application of Polymer Flooding in a Heavy-Oil Reservoir," *J Pet Technol.*, 67, 78–80. https://doi.org/10.2118/0115-0078-JPT

Wilson, K., 2016, "CCEMC Project ID: H110031 H110031" Non-Confidential Final Report, sumitted on Feb. 13. https://eralberta.ca/wp-content/uploads/2017/05/H110031-Nsolv-BEST-Field-Pilot-Final-Report-Public.pdf. Accessed May 13, 2025

Woolford, J.K., Ovalles, C., Fayyaz-Najafi, B., Benson, I.P., Munoz, J.D., 2022, "In situ upgrading of heavy hydrocarbon using one or more degradable solvents and one or more additives", US Patent App. US 2022/0380660 A1.

Yang, Z., Li, X., Chen, H., Ramachandran, H., Shen, Y., Yang, H., Shen, Z., Nong, G., 2019, "Development Optimization for Improving Oil Recovery of Cold Production in a Foamy Extra-Heavy Oil Reservoir", paper SPE- 194628-MS presented at the *SPE Oil and Gas Exhibition* held in Mumbai, India, 9–11 April, https://doi.org/10.2118/194628-MS

Zeynalli, M., Mushtaq, M., Al-Shalabi, E.W., Alfazazi, U., Hassan, A.M., AlAmeri, W., 2023, "A Comprehensive Review of Viscoelastic Polymer Flooding in Sandstone and Carbonate Rocks," *Sci. Rep.*, 13, 17679. https://doi.org/10.1038/s41598-023-44896-9

Zhang, N., Wei, M., Bai, B., 2018, "Comprehensive Review of Worldwide CO_2 Immiscible Flooding." Paper presented at the *SPE Improved Oil Recovery Conference*, Tulsa, Oklahoma, USA, April. https://doi.org/10.2118/190158-MS

Zhou, Y., Yin, D., Chen, W., Liu, B., Zhang, X., 2019, "A Comprehensive Review of Emulsion and Its Field Application for Enhanced Oil Recovery,"*Ener. Sci. Eng.*, 7(4), 1046–1058. https://doi.org/10.1002/ese3.354

3 Thermal Processes

Thermal methods for heavy crude oil (HCO) production involve applying heat to reduce the viscosity of the oil, making it easier to extract from reservoirs in which non-thermal technologies have been exhausted, proven ineffective, or are economically challenged. In general, thermal processes can achieve recovery factors of **30–70%** of the Original Oil in Place (OOIP) (Alvarez 2013).

The most utilized thermal routes are hot water injection, cyclic steam injection (CCI), steamflooding, and Steam-Assisted Gravity Drainage (SAGD). These technologies have been successfully implemented (i.e., Technology Readiness Levels [TRLs] level 9, see **Table 1.2** for definitions) for the Enhanced Oil Recovery (EOR) of HCOs with multi-well and multi-field applications worldwide. However, due to their maturity and high degree of optimization, continuously improving on these routes has a relatively low potential for reducing greenhouse gas (GHG) emissions in the future. Their main challenges are the considerably high energy and water consumption and expensive water treatment needed to generate steam as well as the substantial heat losses and significant GHG emissions (Smalley 2000, Lin 2014, Gordon, 2015, and Rosa 2017).

Rosa and coworkers (2017) examined the forest loss and water use for the world's major oil and oil sand deposits (Rosa 2017). They assessed both the actual and potential rates of water consumption and forest degradation in Canadian deposits, where oil sand extraction is already underway, as well as in other significant HCO deposits globally. Their findings suggest that, based on the then-projected production trends, the exploitation of these resources could demand up to 1.31 km^2 of freshwater per year and result in the loss of approximately 8,700 km^2 of forest. Due to the high water and areal footprint demand, the anticipated increase in HCO extraction could have severe environmental consequences (Rosa, 2017, Ali 2017). Therefore, substantial efforts should be made to mitigate the environmental impacts of HCO and bitumen production while also meeting the energy demands necessary for sustaining global economic prosperity and quality of life (Rosa, 2017).

In line with these goals, over the last two decades, several novel technological approaches have been developed that have significantly reduced GHG emissions, making conventional thermal heavy oil production methods more environmentally friendly while improving oil rates and recovery factors. This chapter will describe relatively recent developments in Low-Carbon Intensity (CI) Steam Injection Technologies (Section 3.1), Steam and Non-Condensable Gases (NCGs) Co-Injection (Section 3.2), Electromagnetic Heating (Radiofrequency [RF]) (Section 3.3), and the Use of Low CI Energy for the production of HCOs (Section 3.4). The potential GHG emission reductions, techno-economic assessments (TEA), life cycle analysis (LCA), and their respective TRL levels will also be examined.

DOI: 10.1201/9781003635475-3

3.1 LOW-CARBON INTENSITY STEAM INJECTION TECHNOLOGIES

Three types of steam-based processes—Cyclic Steam Stimulation (CSS, also known as huff-n-puff or steam soak), steamflooding, and SAGD—have been used alongside solvent co-injection to make the HCO production more environmentally friendly. Since 1980, these processes have been thoroughly studied in the literature and have undergone field testing (Redford 1980, Ardali 2012, Nasr 2001, 2009, Guo 2016, Ovalles 2019, and Ahmadi 2023). Hydrocarbon solvents, like propane, butane, Liquefied Petroleum Gases (LPG), n-pentane, n-hexane, or naphtha, are generally utilized. Other organic solvents have received attention, such as dimethyl ether (DME). They help reduce the viscosity of the heavy oil, making it more mobile. Also, solvents can expand the steam chamber, allowing for a wider area of heated oil to be produced. Thus, higher energy efficiency (EE) and lower GHG emissions are obtained (Fayazi 2019, Wang, 2018; Ahmadi, 2023).

Conversely, the main disadvantage of solvent-based processes is the higher cost of solvents (C5, naphtha, etc.) compared to the produced crude oil. Similarly, the recovery and recycling of the solvent are paramount to increasing the economic benefits. Also, the handling and storage of these materials during steam injection represents an added burden that could make field operation more capital and operating-intensive (Ovalles 2019).

This section will explore CSS-, steamflooding-, and SAGD-solvent co-injection technologies. Environmental aspects and TRLs will be examined as well. The use of additive-steam co-injection methods (i.e., steam/foam and steam/surfactants) offers alternative ways to enhance the efficiency of steam-only processes for HCO production, as well as decreasing the CI of the process. Several examples of these technologies have been documented in the literature, which will be discussed in the final subsection.

3.1.1 Cyclic Steam Stimulation with Solvent Co-Injection

In one of the earliest field tests, Escobar and coworkers reported their experiences of CSS using solvent-based processes in the Bachaquero reservoir, Maracaibo Lake, Venezuela (Escobar 1997). In their study, they applied a slug of a light naphtha solvent (6–7 wt.%) before introducing steam in two of the wells, while in the other nine wells, they utilized a mixture of steam and solvent (3–7 wt.%). Regardless of the method by which the solvent was introduced, the results indicated an average increase of 86% in extra oil production compared to the previous cycle.

Another case study is the Liquid Addition to Steam for Enhancing Recovery (LASER) technology. **Figure 3.1** illustrates the stages of the LASER process as applied to the Clearwater reservoir in Cold Lake, Alberta (Leaute 2002, Leaute 2007, and Stark 2013). This process was developed, patented, and implemented by Imperial Oil as a mid-life technology following CSS. In Stage 1, steam—equivalent to 1,400–1,900 barrels per day (BBL/D) of cold water, containing 5–6% by volume of solvent—is injected into vertical wells at pressures ranging from 1,450 psi (10 kPa·s) to 1,750 psi (12 kPa·s). In Stage 2, the wells are allowed to soak, enabling the

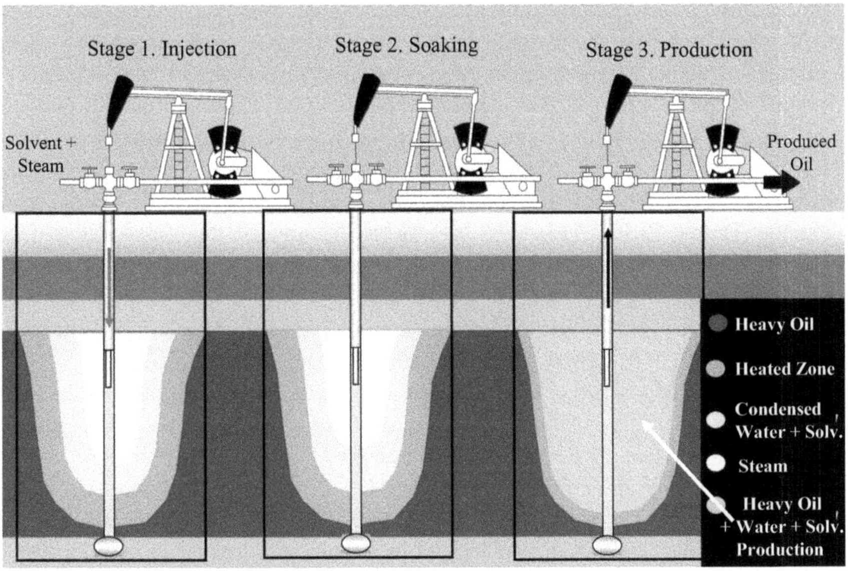

FIGURE 3.1 Stages of Liquid Addition to Steam for Enhancing Recovery (LASER) process (Ovalles 2019).

Adapted from NIPER (1986).

injected heat to penetrate the formation. In Stage 3, production from the wells begins. This process is repeated over multiple cycles, with each well undergoing 8–10 cycles of CSS (Stark 2013).

As seen in **Figure 3.1**, the solvent is introduced into the vapor phase and rises alongside the steam. Stark reported that the production mechanism for the LASER Process is via transportation of the diluent (C5+ condensate) into the deeper parts of the reservoir where steam condensation occurs at the colder boundaries of the formation (Stark 2013). Other researchers have reported similar production mechanism (Nasr 2009, Guo 2016). The solvent then mixes with the HCO, resulting in a reduction of viscosity. Higher bitumen mobility reduces the steam-to-oil ratio (SOR) and reduces the GHG intensity of the process (Imperial 2023). Studies have indicated significant oil production increases of 30–60% compared to conventional thermal projects. However, solvent recovery rates have been reported to be between 80% (Imperial 2023).

Based on laboratory and field observations, the influence of solvent on the recovery mechanism of steam-solvent operations are attributed to the following phenomena: heavy oil viscosity reduction, solution-gas drive throughout pressure drawdown or depletion, modifying absolute and relative permeability caused by the dissolution of organic substances or the mechanical elimination of debris, mobility ratio improvement, interfacial tension reduction, water in oil emulsion breaking, and wettability alteration (Ovalles 2019 and Ahmadi 2023).

Determining the carbon footprint associated with heavy oil recovery using steam-based thermal methods can be challenging. This complexity arises from the various alternatives available for steam generation when compared to traditional oil recovery techniques (Canbaz 2021). Consequently, GHG emissions can vary based on several factors, such as the modeling scope and the types of projects analyzed (for instance, specific projects vs. industry averages), the type of crude oil, the characteristics of the recovery method, and variations in the assumed energy intensities of extraction and upgrading (Brandt 2012). However, it is well-documented that one of the main advantages of steam-solvent co-injection recovery processes for HCO is their reduced energy consumption. This lower energy requirement leads to decreased CO_2 emissions, lower water consumption, and reduces the financial resources needed for water treatment (Wang, 2018; Fayazi, 2019; Ovalles 2019, and Ahmadi, 2023).

Orellana *et al.* conducted a lifecycle assessment associated with the extraction of bitumen using CSS in an Alberta, Canada reservoir (Orellana 2018). The authors found that the median GHG emissions associated with CSS is 77 kilograms of carbon dioxide equivalent removed for one US Barrel of petroleum (kg CO_2e/BBL) bitumen, with a range of 61–109 kg CO_2e/BBL at an 80% confidence level. In a separate study, Suranto and coworkers performed a numerical simulation study using a synthetic reservoir model based on real field data to investigate the effect of solvent on steam injection (Suranto 2015). As before, their findings revealed that a high solvent concentration leads to lower CO_2 emissions and a higher oil recovery factor. They determined that the optimal condition occurs at a solvent concentration of 5.0% v/v. At this concentration, the Net Present Value increased by 21% (equivalent to US$7.76 million), while CO_2 emissions decreased by 9.1% (or 18.05 ton) compared to the scenario using pure steam (Suranto 2015).

Lee and Babadagli conducted a comparative analysis of the GHG emissions generated by steams/solvent processes in comparison with steam-based technologies (Lee 2021). For CSS, the results indicated that the direct GHG emissions ranged 60–100 kg CO_2e/BBL, while for CSS with solvent co-injection varied in the range of 40–70 kg CO_2e/BBL (Lee 2021). Similarly, Leaute reported 25% GHG emission reduction by using Imperial's LASER (Leaute 2007). Thus, a GHG potential emission reduction of medium and a TRL level 9 (multi-well and multi-field applications worldwide, see **Table 1.2**) can be assessed for CSS with solvent co-injection, as illustrated in the high-level technology assessment (**Figure 1.5**).

3.1.2 Steamflooding with Solvent Co-Injection

The co-injection of solvent during steamflooding has been implemented by Shell in a process known as Vertical-well Steam Drive (VSD). Core flooding experiments were carried out using a slug of 15% solvent (C4-C20) (Hedden 2014). Numerical simulations have shown that, in the beginning, bitumen rates are higher in the presence of the solvent than those calculated for the no-solvent case. Recovery improvement in the order of 3–6% of the OOIP was obtained in comparison with the control experiment (Hedden 2014).

The simulations showed that the dominant recovery mechanism is the formation of a solvent bank that mixes the bitumen ahead of the steam zone. Due to this mixing, the viscosity of the bitumen is significantly reduced and thus accelerated oil rates are obtained (Hedden 2014). Also, the solvent bank displaces the HCO and reduces its saturation compared to the case without solvent. Due to the combination of these two effects, an increase in recovery is obtained for the VSD case (Hedden 2014).

The VSD process was piloted in the Peace River area in Canada in two inverted 5-spot 5-acre drive patterns (Castellanos-Diaz 2016). Before steamflooding, CSS cycles were used to create communication between injectors and producers. After several months of the start of the stream drive, a large solvent slug (15% for 6 months) was injected, followed by a steam drive. Higher oil-steam ratios were obtained in the presence of the solvent than those found for the steam-only case (Castellanos-Diaz 2016). After the initial peak, oil rates declined but sustained higher values than those for steam-only patterns for about six months. Field data and reservoir simulation showed that oil rate increases of 27–45% with respect to the no-solvent case, with 32% reduction in the SOR. The authors reported 51–59% of solvent recovery after the trial (Castellanos-Diaz 2016).

Brandt and Unnasch calculated the EE and GHG emissions from thermal EOR for three California-specific cases (Brandt 2010). **Figure 3.2** illustrates the upstream GHG emission in kg CO_2e/BBL for the steamflooding. In the California-specific average case, 60% of steam is assumed to come from Once Through Steam Generator (OTSG) facilities and 40% from Simple Cycle Steam Generation (SCGT) cogeneration facilities, with a statewide-average producing SOR of 5.13 BBL of steam/BBL of oil (Brandt 2010). The low and high California-specific cases presented in **Figure 3.2** were developed using Kern River and Midway-Sunset field

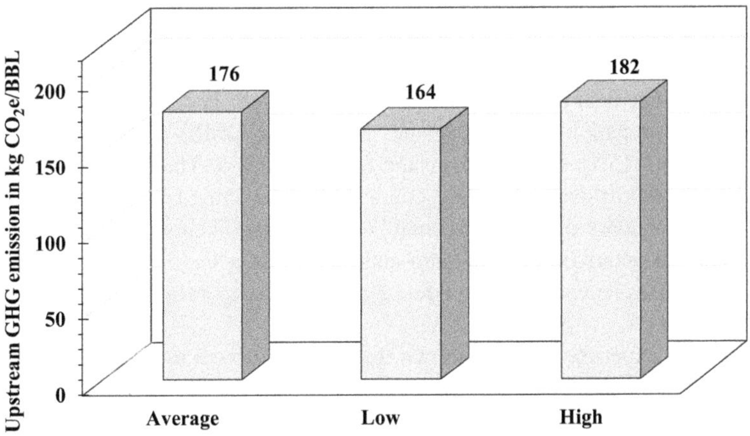

FIGURE 3.2 California upstream GHG emission in kg CO_2e/BBL for the steamflooding Enhanced Oil Recovery (EOR).

Data taken from Brandt and Unnasch (2010).

data. Additionally, the California average case uses the production-weighted average American Petroleum Institute (API) gravity of 17.8°API. In contrast, the California low and high cases use API gravities of crude from Kern River and Midway-Sunset fields of 13.1°API and 18.2°API, respectively (Brandt 2010). The GHG emissions for steam-based EOR ranging 164–182 kg CO_2e/BBL (Brandt 2010), which is significantly higher than the emissions reported for light and medium crude oils of 40 kg CO_2e/BBL, as shown in **Figure 1.3**.

In Venezuela, no successful applications of steamflooding at a field scale have been reported due to the formation of complex emulsions with high viscosity (Rodriguez 2022), greater depths (>1000 m), and the high costs associated with continuous steam generation and injection (Ovalles 2002 and Vega 2011). In contrast, steamflooding with solvent co-injection has been assessed in various locations around the world, including Ikiztepe Field in Turkey (Ahmadi 2023) and the Duri Field in Indonesia (Canbaz, 2021).

According to the literature, the CI for steamflooding range of 53–180 kg CO_2e/BBL, with an average value of 117 kg CO_2e/BBL (Brandt 2010, Gordon 2015, Petroleum 2020, Canbaz 2021, and California 2023). When considering a 32% reduction in SOR observed by Castellanos-Diaz and colleagues during their field test (Castellanos-Diaz 2016), the upstream CI for steamflooding with solvent co-injection is estimated to range 36–122 kg CO_2e/BBL, with an average of 79 kg CO_2e/BBL Based on these findings, the TRL is assessed to be between 7 and 8, indicating limited deployment potential (see **Table 1.2**), along with a low to medium potential for GHG emission reductions (see **Figure 1.5**).

3.1.3 Solvent/Steam-Assisted Gravity Drainage

This process is also known as Expanded Solvent SAGD (ES-SAGD) or Solvent-Assisted SAGD (SA-SAGD) because the solvent is co-injected with steam in a vapor phase. Both terms describe solvent-enhanced SAGD techniques; ES-SAGD is a more specific approach focusing on light hydrocarbon co-injection, whereas SA-SAGD is a general term for solvent/SAGD methods. In this concept, a hydrocarbon solvent (propane, butane, pentane, hexane, or DME) is co-injected at low concentration (<15%) with steam in the SAGD process. The hydrocarbon solvent is selected so that it evaporates and condenses at the same conditions as the water phase at the boundary of the steam chamber. This phenomenon leads to a condensation around the interface of the steam chamber, dilutes the oil, and in conjunction with heat, reduces its viscosity (Ayodele 2009, Bayestehparvin 2018, Nasr 2009, and Ovalles 2019).

Laboratory experiments have shown that adding solvent to steam led to a higher oil rate and reduced the SOR which, in turn, increases the ultimate recovery (Ayodele 2009 and Nasr 2009). **Figure 3.3** shows a mechanistic diagram of the ES-SAGD process and the different regions of the steam chamber during solvent co-injection (Jha 2012). As the temperature drops near the chamber boundary, steam starts condensing first because its mole fraction and partial pressure are higher than that for the solvent (gray zone, **Figure 3.3**). As temperature declines toward

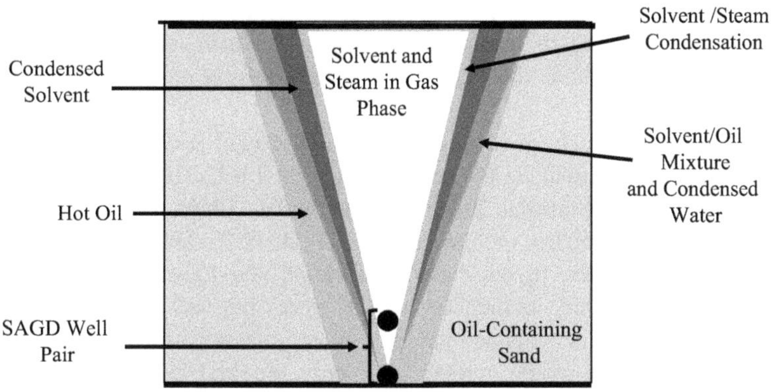

FIGURE 3.3 Mechanistic diagram of the Expanded Solvent SAGD (ES-SAGD) process showing the different regions of the steam chamber during solvent co-injection (Ovalles 2019).

the chamber boundary, the vapor phase becomes increasingly richer in solvent (red region, **Figure 3.3**), and this material and steam condense simultaneously. Thus, contrary to steam-only injection, where condensation occurs at the injected steam temperature, condensation of the steam/solvent mixture is accompanied by a reduction in temperature in the condensation zone and the farther regions (Jha 2012).

The condensed steam/solvent mixture drains outside the chamber, leading to the formation of a mobile liquid stream where heated oil, condensed solvent, and water flow together to the production well (blue region, **Figure 3.3**). The condensed solvent mixes with the heated oil and further reduces its viscosity (Yellow zone, **Figure 3.3**). This additional viscosity reduction more than offsets the effect of reduced temperature near the chamber boundary (Jha 2012). As the steam chamber expands laterally because of continued injection and temperature increase, the condensed solvent mixed with oil, which in turn lowers the residual oil saturation in the steam chamber. According to Hascakir, solvent co-injection with steam reduces the interfacial forces between asphaltenes and water, leading to a lower water-in-oil emulsion percentage than steam injection alone (Hascakir 2016). Therefore, ultimate oil recovery with the steam/solvent-co-injection process is higher than that in the steam-only counterpart (Jha 2012).

Furthermore, Ivory and coworkers proposed a hybrid EOR process called *In-Situ* Reflex (ISR), also known as the thermal solvent reflux process, which utilizes resistive electric heaters or steam injection to vaporize the solvent and recycle mobilized water downhole (Ivory 2010, Sadeghi 2022, and Zeidani 2024). The authors noted that the additional heat serves to initiate communication between the injector and the producer. Additionally, it will increase the rate of solvent diffusion into the oil. Results indicated that ISR delivers oil rates comparable to SAGD, but with lower energy input, resulting in concomitant reductions in GHG emissions. In laboratory experiments using Hillmond oil, both electrically heated wells and steam-solvent hybrid setups, produced similar oil rates and recovery factors. The ISR process is designed to recycle solvent *in situ*, which enhances recovery and economic viability without the need for continuous solvent injection (Ivory 2010 and Zeidani 2024).

Table 3.1 presents a summary of several field tests of solvent co-injection with SAGD (Gupta 2005, Gupta 2006, Orr 2009, Alberta 2014, 2015, 2016a, and 2016b, and Connacher 2023, Surmont 2024). For the Senlac, Christina Lake, Long Lake, Surmont, and Algar projects, increases in the oil rate (BBL/D) in the 30–40% range were reported. However, in the cases of Jackfish and Leismer, the results were inconclusive. Additionally, reductions of SOR were observed in the 20–40% range.

Jaimes and coworkers evaluated the intermittent injection of a solvent, along with steam, aimed at reducing the amount of energy required to extract oil from the Canadian oil sands (Jaimes 2019). The authors conducted numerical simulations to assess the effects of the type of solvent, cycle duration, solvent concentration, and the number of cycles over 10 years. Methane, ethane, propane, butane, pentane, hexane, and CO_2 were considered under varying injection schedules. The results revealed that SAGD, combined with intermittent injection of hexane, resulted in a 45% increase in cumulative oil production and a 23% increase in the cumulative energy-oil ratio (Jaimes 2019). Based on the above-presented precedents, it can be concluded that the solvent co-injection SAGD process decreases energy requirements, reduces steam usage, lowers GHG emissions, and minimizes water consumption. Moreover, the percentages of solvent recovery range from 20 to 80% (**Table 3.1**). Several reasons have been proposed, such as loss to the mineral formation, pipeline leaks, difficulties in sampling operations, and solvent composition overlap with the bitumen, making it difficult to quantify (Alberta 2018).

According to Fassihi and Kovscek, the SOR is a key metric for evaluating the efficiency of steam injection processes (Fassihi 2017, p. 181). Generally, a higher SOR is linked to increased GHG intensity, as it indicates lower recovery efficiency. In the case of SAGD, an increase in the SOR from 0.2 to 0.4 leads to a 4-fold rise in GHG emissions, as measured in kg CO2e/BBL. Fassihi also noted that integrating electrical power generation with steam production—commonly referred to as cogeneration—can reduce net GHG intensity by approximately 20% (Fassihi 2017). Similar reductions in GHG emissions are anticipated when using solvent-assisted SAGD technology.

Orellana *et al.* (2018) conducted a lifecycle assessment of bitumen extraction using SAGD in Alberta, Canada. The authors found that the median GHG emissions were 68 kg CO_2e/BBL, with a range between 49 and 102 kg CO_2e/BBL at an 80% confidence level. They reported that the median emissions intensity of Alberta's SAGD projects has remained relatively stable from 2000 to 2013, despite a more than 6-fold increase in heavy oil and bitumen production. Variability between projects is the largest source of emissions, largely driven by reservoir characteristics and operational issues (Orellana 2018).

Figure 3.4 shows the average upstream CI (in kg CO_2e/BBL) for different HCO production technologies such as Vapor Extraction Process (VAPEX), Nsolv, SAGD, Solvent/SAGD, Solvent/CSS, and CSS, as published in the literature by different authors (Orellana 2018, Hoque 2023, Soiket 2019, and Lee 2021). As anticipated,

TABLE 3.1
Summary of Selected Field Tests of Solvent Co-Injection with Steam Assistance Gravity Drainage (SAGD) for Heavy Crude Oil (HCO) Production[a]

Project (Years)	Operator	Solvent Used[b] (% v/v vs. Steam)	Oil Rate (BBL/D)[c]	Steam-to-Oil Ratio (SOR)[d]	Solvent Recovery[e]	References
Senlac (2002)	Encana	Butane (15 wt.%)	From 1,900 BBL/D to 3,000 BBL/D (~58% inconclusive)	From 2.6 to 1.6	~70%	Gupta (2005)
Christina Lake (2004)	Encana	Butane (15 wt.%)	From 630 BBL/D to 1900 BBL/D (~58% inconclusive)	From 5 to 1	NA	Gupta (2006)
Long Lake (2006)	Nexen	Jet B (C7-C12) (5%)	Same as SAGD	Same as SAGD	NA	Orr (2009)
Jackfish (2013)	Devon	C6	Inconclusive	Inconclusive	NA	Alberta (2014)
Leismer (2013–2015)	Statoil	C5-C6 (average 5%)	Non-responsive	From 1.5 to 1.0	20% (Q2 2015)	Alberta (2016b)
Algar (2011–2023)	Connacher	C5-C6 (5–10%)	Increase of 20–30%	Reduction of 28–32%	70% (2022)	Alberta (2016a), Connacher (2023)
Surmont (2012–2024)	Conoco-Phillips	C3-C4-C6 (20%)	Increase of 20–30%	Reduction of 10–30%	50% (February 2016)	Alberta (2015), Surmont (2024)

[a] NA: Not available or not reported.
[b] Percentage by v/v of solvent vs. cold water equivalent of steam.
[c] Oil rate produced in barrels per day. The number in brackets indicate the percentage of increase vs. SAGD.
[d] Cold water equivalent of steam vs. oil produced (v/v).
[e] Percentage of solvent recovered vs. volume injected. Numbers in bracket indicate the time of measurement.

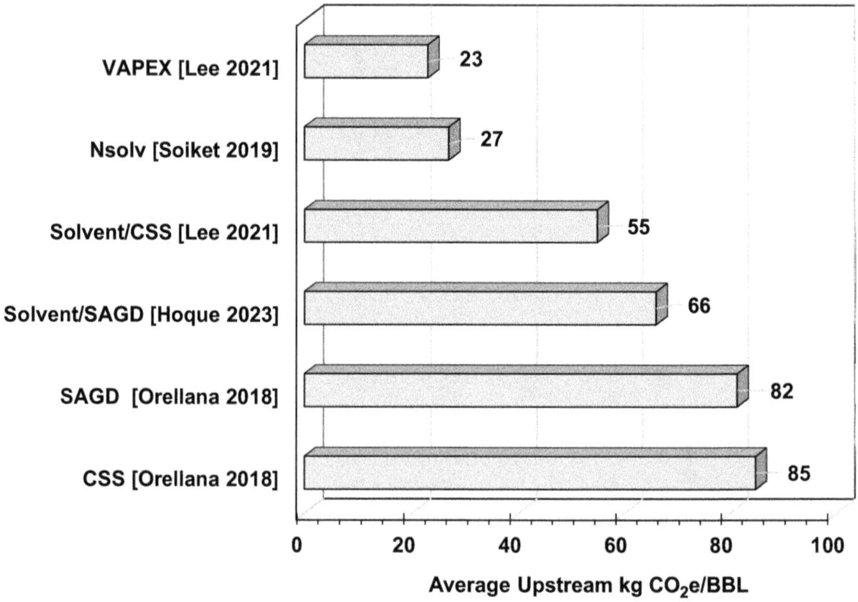

FIGURE 3.4 Average upstream carbon intensity (CI) in kg CO_2e/BBL for different heavy crude oil (HCO) production technologies. Abbreviations: VAPEX, Vapor Extraction Process; Nsolv, Solvent Deasphalted Process; SAGD, Steam-Assisted Gravity Drainage; CSS, Cyclic Steam Stimulation.

solvent-based technologies exhibit lower carbon intensities than their steam-containing counterparts. The use of solvents in CSS and SAGD operations resulted in a ~35 and ~20% reduction in GHG emissions, respectively. CSS and SAGD appear to be the least environmentally friendly technologies, with CI's of 85 and 82 kg CO_2e/BBL, respectively (**Figure 3.4**). This topic will be discussed in detail in Section 5.1. Based on these results, a GHG potential emission reduction of medium and a TRL level 9 (multi-well and multi-field applications worldwide, see **Table 1.2**) can be assessed for Solvent/SAGD co-injection, as illustrated in the technology assessment shown in **Figure 1.5**.

3.1.4 Steam/Additive Co-Injection

Developments in the use of steam/additive co-injection methods (i.e., steam/foam and steam/surfactants) offer alternative ways to enhance the efficiency of steam-only processes for HCO production, as well as decreasing the CI, water usage, and aerial footprint. The steam/foam and steam/surfactants improvement in oil production is based on the mechanisms of reduction of the surface tension between oil and water, increasing the viscosity of the injected liquid, changes in wettability (making rock surface more water-wet), and improving the sweeping efficiency of the injected liquid (Fassihi 2017, Lee 2021, Rodriguez 2022, and Ahmadi 2023). However, there are several limitations of the additive-steam co-injection, such as the potential for formation damage, the cost of the chemical additives, and the complexity of the injection process (Alsabagh 2023).

In the case of steam/foam applications, the main objective is to lower the SOR by creating foam from the steam using a surfactant solution. As foam is defined as a dispersion of a gas within a continuous liquid phase, the surfactant molecules position themselves at the gas-liquid interface due to thermodynamic principles. The repulsion between these surfactant molecules helps stabilize the thin liquid films that separate the gas bubbles, preventing them from coalescing. By dispersing the gas phase, gas mobility is significantly reduced. Both laboratory and field testing have shown that steam-foams have mobility reduction factors (gas mobility divided by steam-foam mobility) in the range of 10–100 (Fassihi 2017). Several types of surfactants have been used as steam-foaming agents, such as linear toluene and xylene sulfonates, carboxylates, sodium lauryl sulfate, alpha olefin sulfonates, as well as alkali co-injection (Lee 2021 and Ahmadi 2023).

For steam and surfactant applications, the additives should function as detergents to alter the wettability of the formation from oil-wet to water-wet. This change, combined with a reduction in interfacial tension, allows water to displace the reservoir more easily, leading to an increase in oil production. Several additives have been investigated, including deep eutectic solvents, ionic liquids, nanofluids, linear toluene sulfonates, and olefin sulfonates (Lee, 2021). Thermal stability is a crucial factor in selecting additives (Odier, 2021). For this reason, alkyl sulfonates are commonly used due to their excellent resistance to heat and pH changes (Taylor, 2018).

Steam/additive co-injection technologies have been developed since the late 1980s. **Table 3.2** summarizes selected field tests conducted in various locations around the world from the early 1990s to the early 2020s. Each case study included in **Table 3.2** highlights a different aspect of the technology and demonstrates the effectiveness of steam/foam and steam/surfactant processes in enhancing the production of HCOs.

In one of the earliest tests in 1993, Gauglitz and coworkers from Chevron (**Table 3.2**, entry 1) conducted a steam-foam trial in Midway-Sunset 26C (California, USA) to optimize the use of steam-foam for improving profile control in a vertical steam injection well without the addition of a NCG (Gauglitz 1993). The field experiments were designed to find the optimal surfactant concentration and liquid volume fraction (LVF), or steam quality, for improving injection profiles with steam-foam. Radioactive tracer measurements were conducted with and without foam injection. Steam-foam co-injection enhanced the injection profile for concentrations exceeding 0.1 wt.% of Chaser™ SDI020 at a steam LVF of 0.02 for the well tested. These tests fully demonstrated that profile improvement can be achieved with surfactant injected at about half the rate that was previously known as necessary for in-depth foam applications, and this was accomplished without the addition of nitrogen (Gauglitz 1993).

In 1994, Friedmann and coworkers from Chevron (**Table 3.2**, entry 2) studied the mechanism of steam-foam co-injection in the presence of nitrogen (NCG) in a seven-well pattern of Section 26C of the Midway-Sunset field (California, USA). Their goal was also to measure the incremental oil that can be attributed to the formation of

TABLE 3.2
Case Studies for Steam/Additive Co-Injection for Heavy Crude Oil (HCO) Production[a]

S. No.	Field (Years)	Operator (Country)	Thermal Process[b] (Additive Injection)	Additive[c] (Type)	Additive Concentration[d]	Main Finding/Results	References
1	Midway-Sunset (1993)	Chevron (California, USA)	Continued (co-injection)	Chaser SD1020 (foaming agent)	0.085–0.36 wt.%	Using radioactive tracers, improved injection profiles vs. no additive	Gauglitz (1993)
2	Midway-Sunset (1994)	Chevron (California, USA)	Steamflooding (co-injection)	Chaser SD1000 (foaming agent)	0.1–1.0 wt.%	Pattern oil rate increased from 350–422 BBL/D (20% incremental)	Friedmann (1994)
3	Bolivar Coast (1992)	PDVSA (Venezuela)	CSS (co-injection)	LTS and BBS (foaming agent)	0.25–2 wt.%	LTS is more efficient than BBS. Increments in cumulative oil vs. no additive	Mendez (1992)
4	Bachaquero (1999)	PDVSA (Venezuela)	CSS (co-injection)	LAAS (foaming agent)	0.75–0.85 wt.%	For vert. wells, 15–40% more cumulative oil per cycle vs. no additive	Valera (1999)
5	Athabasca (1991)	Amoco/Chevron (Canada)	Steamflooding (co-injection)	Chaser SD1020 (foaming agent)	NA	Broadness of temperature/profiles. Average oil rates higher vs. no additive	Sander (1991)
6	Shengli (2014)	China University of Petroleum (China)	CSS (co-injection)	Tannin gel (foaming and FRC-2 surfactant)	0.2%	Pressure increased after injection. Cumulative oil improved, and the water cut decreased	Li (2014)
7	Magdalena Valley (2021)	Ecopetrol (Colombia)	CSS (pre-formed)	Foaming agent with N_2	0.25 wt.%	Higher steam injection pressures. Incremental oil recoveries > 20 MBBL	Pérez (2022)
8	Athabasca (2022)	Conoco-Phillips (Canada)	SAGD (co-injection)	NA (surfactant)	0.3 wt.%	Reduction in the produced emulsion viscosities. GHG intensity reduction of 20%	Sanchez (2022)

[a] NA: Not available or not reported.
[b] CSS: Cyclic Steam Stimulation, SAGD: Steam-Assisted Gravity Drainage.
[c] LTS: Linear sodium alkyl toluene sulfonate, BBS: Branched alkylbenzene sodium sulfonate, LAAS: Linear ionic-alkyl-aryl sulfonate. Chaser™ SD1000 and SD1020: Proprietary sulfonate additive. FRC-2: High temperature anionic surfactant.
[d] Additive concentration where wt.% = Weight percentage (w/w).

the foam. They continuously injected surfactant along with steam and nitrogen, which resulted in an increase in the bottomhole injection pressure from 689 kPa (100 psi) to 2,068 kPa (300 psi), indicating foam generation. Improvements in both vertical and areal sweep efficiency of steam were also observed. The authors reported that oil production increased from an average of 350 BBL/D before foam injection to 422 BBL/D. This represents an average increase of 20% (Friedmann 1994). In a recent report, the CI for CSS/additives in California was estimated to be more than 60% lower than that found for the conventional CSS, dropping from 106 to 40 kg CO_2e/BBL (Clean 2023).

After conducting laboratory experiments with steam/foam, Mendez *et al.* from PDVSA conducted field evaluations (**Table 3.2**, entry 3) in a set of 16 wells using two commercial foaming agents at 0.25–2 wt.% (LTS: Linear sodium alkyl toluene sulfonate, BBS: Branched alkylbenzene sodium sulfonate) (Mendez 1992). The field tests were performed in the Tia Juana, Lagunillas, and Bachaquero fields on the Bolivar Coast, Venezuela. These reservoirs contain crude oil with an average API of 11°, a dead oil viscosity of 20 Pa·s (20,000 cP) at 43°C (110°F), and a pay zone thickness of 9–34 m (30–110 ft). In the case of the LTS surfactant, wells that exhibited higher foam's blocking efficiency also showed the largest incremental increases in cumulative oil production during the fourth cycle of the CSS, compared to model predictions without the foaming agent. Conversely, BBS demonstrated a different behavior: the additional oil production was not directly proportional to the foam's blocking efficiency, and in some wells, there was no observed increase in production at all (Mendez 1992).

Valera and coworkers from PDVSA (**Table 3.2**, entry 4) performed steam-foam field tests in vertical and horizontal wells of the Bachaquero field located in Maracaibo Lake Basin, Venezuela (Valera 1999). This reservoir has an 11.5°API crude, a payzone thickness of 24 – 91 m (80–300 ft), an area of 23,511 acres, and 6,621 million BBL of OOIP. Most wells suffered from lower than expected production as a consequence of gravity override and steam channeling during CSS. Linear ionic-alkyl-aryl sulfonate (LAAS) was used as a foaming agent to block the most permeable areas and achieve a uniform injection profile. The authors found 15–40 and 83% more cumulative oil per cycle vs. no additive for vertical and horizontal wells, respectively (Valera 1999). These results, along with the previous findings (**Table 3.2**, entry 3), confirm the effectiveness of foaming agents as an EOR process for the exploitation of Venezuelan heavy oils. Unfortunately, no results were reported for the SORs or CI reductions.

A team from Amoco and Chevron (**Table 3.2** entry 5) performed the first steam-foam flooding in the Gregoire Lake *In-Situ* Steam Pilot of the Athabasca field (Alberta, Canada) (Sander 1991). Chaser™ SDI020 (surfactant) was used to mitigate steam override and increase oil production. Additional testing was performed to optimize surfactant and steam injection rates. Foam generation was demonstrated by increases in the injection well pressure, and foam propagation was evidenced by the broadening of the observation well temperature profiles. Steam diverted by the foam caused heating of the previously bypassed tar-rich section of the reservoir and resulted in a doubling of oil production. The average oil rate was higher (~20–30%) than the rate following the steam-only stimulation (Sander 1991).

Li *et al.* from China University of Petroleum (**Table 3.2** entry 6) investigated the performance of tannin foam and anionic surfactants for steam-injection profile modification in CSS (Li 2014). They conducted laboratory experiments using single and parallel sandpacks under high temperatures to determine the effects of foam quality, permeability, injection mode, temperature, permeability ratio, and oil saturation on blocking and profile modification capabilities of tannin foam. Then, field tests were conducted in the D41 Block in Shengli Oilfield in China. The crude oil viscosity ranged from 320,000–560,000 mPa·s under reservoir conditions. The depth of the formation was 1,320 m, and the average permeability of the formation was 2,400 milli Darcy. Four-to-five CSS cycles were carried out in two steam-stimulated wells before tannin foam injection. At the end of the final cycle, the water cut exceeded 95%, and there was minimal oil production. Tannin foam was then applied to enhance the steam profile, resulting in observed increases in wellhead pressure of up to 2 MPa. This suggests that the tannin foam effectively blocked steam channels in the formation. The authors reported that cumulative oil production for the two CSS wells increased by 400–750 ton (2,500–4,700 BBL), while the water cut decreased by 7–15% (Li 2014).

Pérez and coworkers from Ecopetrol (**Table 3.2**, entry 7) conducted six field tests of pre-formed foam injection prior to a steam cycle to optimize CSS processes in mature wells (>10 cycles) in a heavy oilfield located in the Middle Magdalena Valley, Colombia (Pérez, 2022). Their methodology involved mixing a foaming agent with nitrogen, as a NCG, using a specially designed wellhead. Next, the foamed mixture is co-injected with a steam downhole. The first two pilot wells demonstrated incremental oil recoveries exceeding 20,000 BBL of cumulative oil. The authors reported that the other four wells were producing above the CSS baseline. Higher steam injection pressures were observed, indicating a shift in flow diversion to lower permeabilities and non-swept oil intervals. The latter was confirmed by geochemical analysis, which revealed an increase in the total concentration of water-soluble organic compounds specific to the non-swept zone (Pérez, 2022).

Pérez *et al.* (2022) determined the EE and CI indexes for the steam-foam co-injection and compared them with the steam-only case (Pérez 2022). The EE index (no units) was calculated by dividing the energy consumed by the energy produced. When the EE index values were below one (EE < 1), it indicated that the process is energy efficient, producing more energy (e.g., oil) than the energy required (e.g., steam, additives, equipment, etc.). Conversely, an EE value greater than one suggested inefficiency, meaning that the process consumes more energy during EOR than the oil produced. Results from a 27-month steam cycle demonstrated that the hybrid technology had higher EE compared to the conventional CSS (0.08 vs. 0.2, respectively) (Pérez 2022). The CSS/foam co-injection also showed a lower CI (~45 kg CO_2e/BBL) than that calculated for CSS (~94 kg CO_2e/BBL). The authors noted that both technologies were inefficient at the early stages of the production cycle (<7 months) due to high-water production and improved as the production cycle advanced. As mentioned, as time progressed, CSS/foam co-injection outperformed steam-only CSS (Pérez 2022). The author concluded that extending the CSS production cycles through the injection of pre-formed foams resulted in greater oil production while using the same amount of natural gas for steam generation. This reduction in natural gas consumption contributed to a lower carbon footprint for generating one barrel of crude oil (Pérez 2022).

Similarly, Osma *et al.* (2019) conducted a numerical simulation of a radial model representing a large heavy oil reservoir in the Middle Magdalena Basin, Colombia. The aim was to determine the EE index along with several economic indicators for various CSS methods, both with and without additives such as foam, solvent, or flue gas. The results indicated that CSS with foam had an EE index of 0.005, making it significantly more efficient than CSS with solvent (1.969), CSS with flue gas (0.211), and the method without any additives (0.9) (Osma 2019).

Sanchez from ConocoPhillips (**Table 3.2**, entry 8) reported a field pilot of steam/surfactant using commercially available chemical additives to improve the performance of SAGD operations by increasing oil drainage rates, reducing produced emulsion viscosities, and minimizing energy losses (Sanchez 2022). Field pilot execution was conducted in four well pairs tied to existing facilities at the Surmont SAGD facility. Overall, the co-injection of steam additives in the pilot wells was achieved over 21 months, from the project's start to its completion date. The results demonstrated improved thermal efficiency, incremental oil production, and a 20% reduction in GHG emissions intensity. This reduction translates to a decrease of 15 kg CO_2e/BBL with respect to conventional SAGD, or approximately 20,000 ton CO_2e/year. The author concluded that the success of the pilot in addressing remaining uncertainties advanced the technology to TRL level 9 (see **Table 1.2**) (Sanchez 2022). This conclusion is in agreement with the technology assessment illustrated in **Figure 1.5**, which corresponds to a potential GHG reduction of medium (GHG reduction between 30 and 70%).

3.2 STEAM AND NON-CONDENSABLE GASES CO-INJECTION

Currently, steam and NCG co-injection is a common practice employed by operators across many SAGD projects (Fassihi 2017, Austin-Adigio 2019, and Kim 2025). This process is also known as Steam and Gas Push (SAGP). It has been reported that about half of all Canadian SAGD well pads are steam/NCG co-injection technologies (Hartley 2025). Nitrogen and methane are commonly used NCGs due to their low solubility in oil; however, concentrations above 3 mol% typically decrease SAGP efficiency (Kim 2025). Steam /NCG co-injection not only enhances the economic viability and EE of heavy oil production projects but also enables operators to redirect valuable steam toward more profitable and efficient lower SOR wells. This practice results in increased oil production, reduced environmental impact, and lower water usage.

Simulations and field test results have demonstrated that substituting steam with NCG lowers the amount of steam injected, resulting in a consequent reduction of the steam's saturation pressure and temperature (Butler 1999 and Austin-Adigio 2019). The NCG functions as an insulation barrier at the top of the reservoir, which helps minimize heat loss to the overburden (Fassihi, 2017). To avoid interference from NCG with the steam chamber, it has been recommended to maintain an injection pressure between 0.95 and 1.1 times the reservoir pressure (Kim 2025). Additionally, thicker thief zones contribute to increased heat loss, while higher vertical permeability accelerates the rise of the steam chamber, necessitating earlier NCG injection (Kim 2025).

Conversely, injecting less steam has a negative impact on the amount of mobilized bitumen. However, numerical simulations conducted by Austin-Adigio and Gates indicate that this impact is relatively minor, and reducing the amount of injected steam can improve the SOR (Austin-Adigio 2019).

Despite its advantages, co-injecting steam with NCG has associated costs. In 2022, over 250 million cubic feet per day (MMcf/d) of natural gas was injected into Canadian SAGD operations, with a significant portion remaining in the reservoir and not being recovered. The percentage of injected gas that returns to the surface is called the recycle ratio, and it varies widely, ranging from approximately 90% to nearly negligible. Assuming a mid-range recycle ratio of 50% and a US$3/Mcf natural gas price, the annual cost of NCG co-injection across the industry would be at least US$137 million (Hartley 2025).

This section will explore four types of steam/NCG co-injection technologies for heavy oil production with TRL levels between 7 and 8 (i.e., limited deployment, see **Table 1.2**) that have medium to high potential for reducing GHG emissions (see **Figure 1.5**). The technologies discussed include steam/natural gas co-injection, Enhanced Modified Steam and Gas Push (eMSAGP), Enhanced Modified Vapor Extraction (eMVAPEX), and steam/flue gas co-injection field tests. Field testing and environmental aspects will be examined as well.

3.2.1 Steam/Natural Gas Co-Injection Field Tests

In the last decade, steam/natural gas (as NCG) co-injection has been field tested in the McMurray and Cold Lake reservoirs of Alberta, Canada. **Table 3.3** presents five case studies that show the effectiveness of this process for the EOR HCOs. Since 2007, JACOS has conducted natural gas co-injection as part of the Hangingstone Demonstration Project (**Table 3.3**, Entry 1) (Doan 2014). They found that bitumen production remained unaffected, while the SOR decreased from 4.1 to 3.4, representing a 17% reduction during the co-injection process. Moreover, there was a reduction of 46 ton in CO_2 emission intensity from the combustion of natural gas per unit of production, equivalent to a 6.4% decrease. The steam injected was balanced against the water produced, and any excess steam was allocated to new wells or non-thermally matured wells to optimize overall field production. (Doan 2014).

Suncor initiated steam/NCG (i.e., methane fuel gas) co-injection on nine mature pads in the Firebag project as of May 2018 (**Table 3.3**, Entry 2) (Suncor 2021 and Kim 2025). In this project, the flue gas and steam co-injection typically commenced on a pad when estimated oil recovery exceeds 65% through normal SAGD operations. The concentrations for injected NCG gas were adjusted in the 0.25–0.42 mol % range. This approach resulted in a reduced SOR, while the overall steam injection at the field level was optimized by reallocating it to less mature SAGD pads, as the previous case (**Table 3.3**, Entry 1). No significant temperature reductions were observed within existing steam chambers from observation well temperature/pressure data (Suncor 2021 and Kim 2025).

TABLE 3.3

Selected Case Studies for Steam/NCG (Natural Gas) Co-Injection Field Tests for Heavy Crude Oil (HCO) Production[a]

S. No.	Field (Years)	Operator (Project)	Thermal Process[b]	Injection Conditions	Main Finding/Results[c]	References
1	McMurray (2007–2013)	JACOS (Hangingstone)	SAGD/natural gas co-injection	5 well-pairs at ~1 mol% natural gas	• Steam/natural gas SOR 3.4 vs. 4.1 steam only (17% reduction) • Reduction of CO_2 emitted of 46 ton/d	Doan (2014)
2	McMurray (2018–2023)	Suncor (Firebag)	SAGD/natural gas co-injection	11 pads at 0.25–0.42 mol% natural gas	• SOR has been reduced • 40% of injection NCG is produced back • No significant temperature reduction	Suncor (2021)
3	Cold Lake (2023)	Imperial (Maskwa)	Steamflooding natural gas co-injection	One pad at average 1.6 mol%	• Natural gas recycle rate was 60 and 70%. • SOR and GHG reductions	Imperial (2023)
4	McMurray (2022–2023)	Connacher (Algar)	SAGD/natural gas co-injection	4 pads at 3–5 mol%	• Improved SORs with minimal impact on production • ~1,600 ton/d steam re-allocated	Connacher (2023)
5	McMurray (2019–2024)	Cenovus (Foster Creek)	SAGD/natural gas co-injection	50 ton/d; 100 mol% during blowdown	• Natural gas (NCG) co-injection as an enhanced recovery late life strategy • Results are promising	Cenovus (2024)

[a] NCG: Non-condensable gas used was natural gas (mostly methane) for all cases.

[b] Thermal production method where SAGD/natural gas co-injection or steamflooding/natural gas co-injection.

[c] SOR: steam-to-oil ratio.

Imperial Oilfield tested the co-injection of NCG with steam in a steamflood area in the Maskwa project, located within the Cold Lake field, Alberta, Canada (**Table 3.3**, Entry 3) (Imperial 2023). The objective was to maintain reservoir pressure and create a gas blanket at the top of the reservoir to reduce heat loss to the overburden. One pad was utilized with an average concentration of 1.6% mole fraction of NCG injected. The operator reported a natural gas recycle in the 60–70% range as well as SOR and GHG reductions (Imperial 2023). Connacher also evaluated the SAGD/natural gas co-injection in the McMurray field utilizing 4 pads with 3–5 mol% of NCG (**Table 3.3**, Entry 4) (Connacher 2023). As before, the results showed improved SORs with minimal impact on production. Also, ~1,600 ton/d of steam was reallocated to other parts of the field to enhance EE (Connacher 2023). Cenovus evaluated the steam/methane co-injection in the Foster Creek project, which is also in the McMurray field (**Table 3.3**, Entry 5) (Cenovus 2024). The strategy was to increase oil production by using NCG (CH_4) co-injection late in the life of the project. The operator reported that results are promising so far (Cenovus 2024).

In 2017, a pilot project in Canada explored the effectiveness of co-injecting an NCG, such as methane, alongside steam (ConocoPhillips 2017). This approach aimed to create an insulating blanket that could minimize potential thermal and pressure losses within the steam chamber. The field results showed a 20–25% reduction in GHG emissions intensity, enhanced cost competitiveness, and promising economic returns (ConocoPhillips 2017).

3.2.2 Enhanced Modified Steam and Gas Push (eMSAGP)

This process was patented by MEG Energy Corp. and involves co-injecting a NCG, like natural gas, with the steam (Yee 2018). Two SAGD well pairs are drilled in the pay zone, as illustrated in **Figure 3.5**. Steam and the NCG are co-injected into the

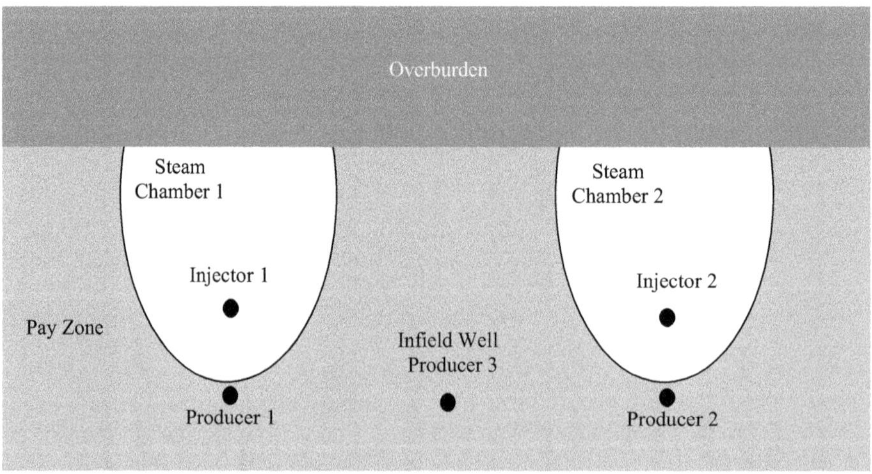

FIGURE 3.5 Simplified lateral view of the Enhanced Modified Steam and Gas Push (eMSAGP) Process.

reservoir through the upper wells (injectors 1 and 2) of the well pairs (Butler 2022 and Yee 2018). The replacement of steam with NCG results in a reduction of the steam injected, leading to a consequent decrease in steam saturation pressure and temperature. The interplay between NCG and steam at the edge of the chambers influences heat transfer, leading to reduced heat transfer to the overburden and sides of the depletion chamber (Austin-Adigio-2019).

After recovering ~30% of the OOIP using Producers 1 and 2, an infield well producer (Producer 3, see the center of **Figure 3.5**) is utilized to collect bitumen that would otherwise be unrecoverable, thereby increasing the overall recovery percentage. The infill well production in eMSAGP shortened the distance from the warm bitumen front to the producer, expanded the swept volume, and reduced residual oil saturation (Zhang 2021). It is reported that by introducing an infill well and operating it in a flooding manner, eMSAGP effectively addressed the uneven development of adjacent steam chambers (steam chambers 1 and 2, **Figure 3.5**), which is a common challenge encountered in field operations (Zhang 2021).

The eMSAGP process has been field-tested at the Christina Lake Project in Alberta, Canada, since 2008 (MEG 2022). This reservoir has an average pay height of 24 m (79 ft), porosity ranging from 0.3 to 0.36, and horizontal and vertical permeabilities of 5 and 2.5 Darcy, respectively. The average temperature is 13°C (55°F), and the initial pressure is 2,100 kPa (304 psi). MEG's operations began with a conventional SAGD production system, which operated with an instantaneous steam-to-oil ratio (iSOR) in the range of 2.25 to 3.25. After the unrecovered bitumen between the SAGD well pairs was heated, the eMSAGP co-injection process commenced, introducing approximately 5 mole% of NCG in the steam. This process resulted in an average reduction in the SOR from 2.75 to 1.50, representing a 45% decrease. Late-life wells are converted to NCG injection-only operation to improve the iSOR further (MEG 2022). MEG reported that the ability to free up steam capacity from operating wells to new wells reduces the capital requirements for increasing production. Approximately 75–80% of the co-injected NCG is recovered, and no negative impacts on well integrity have been observed (MEG 2022).

A LCA was conducted for the eMSAGP process, and the results were compared with those of conventional SAGD (Chen 2019 and Sleep 2021). Two life cycle models were utilized, the Oil Production Greenhouse Gas Emissions Estimator and the Petroleum Refinery Life Cycle Inventory Model, to calculate the GHG emissions. The findings indicated that the commercial application of eMSAGP has a CI of 58 kg CO_2e/BBL. This result represents a 27.5% reduction compared to SAGD (~80 kg CO_2e/BBL) in the same reservoir (Chen 2019 and Sleep 2021).

As discussed, eMSAGP effectively reduces steam demand, leading to lower natural gas consumption and decreased water usage. Since steam generation is the most significant contributor to emissions in SAGD, it can be concluded that eMSAGP lowers CI by optimizing steam utilization. Reduced steam usage also decreases the need for water treatment and disposal, resulting in a smaller environmental footprint. The technology has enabled MEG Energy Corp. to increase production without requiring an increase in steam generation capacity. As shown in **Figure 1.5**, a TRL levels between 7 and 8 (i.e., limited adoption, see **Table 1.2** for definitions) can be assigned a medium GHG reduction potential.

3.2.3 Enhanced Modified Vapor Extraction (eMVAPEX)

The eMVAPEX is a modification of VAPEX (see Section 2.5.1) and was patented by MEG Energy Corp. (Yee 2018). The eMVAPEX concept is an extension of eMSAGP technology (see **Figure 3.5**) and involves the injection of light condensable hydrocarbons (e.g., propane or butane) instead of steam/NCG after the initial SAGD operation has reached a bitumen recovery of approximately 20–30% of the OOIP (MEG 2020 and 2023).

The field test was conducted in one well pair at the Christina Lake project in 2014 by injecting commercial-grade propane at a rate not exceeding 200 m^3/d (standard liquid volume) per well (MEG 2023). The pilot was extended to three more pairs in 2016. In 2018, the propane injection facility was further expanded, resulting in the addition of nine more well pairs utilizing the eMVAPEX technology (MEG 2023). It was found that the injection of condensable gas helped sustain reservoir pressure, allowing for near-zero steam injection with a concomitant reduction in the SOR from 2 to 3 (SAGD) to an average of 0–1.5 (eMVAPEX) (MEG 2020 and 2023). Similar to the eMSAGP technology, the steam that would have been used in the pilot wells was redirected to new SAGD well pairs. Additionally, the eMVAPEX process diluted warm bitumen, enhancing its flow properties and partially de-asphalting the product, which improved the API gravity, thereby increasing overall production efficiency. The process achieves >70% solvent recovery, reducing operational costs compared to some solvent-based methods that experience higher losses (MEG 2020 and 2023).

MEG Energy Corp reported that by replacing steam with propane injection, eMVAPEX reduces the amount of steam required per barrel of bitumen, cutting GHG emissions from 80.6 kg CO_2e/BBL (SAGD) to 45.9 kg CO_2e/BBL (eMVAPEX), which translates to a 43% reduction (MEG 2024). Moreover, traditional SAGD methods require extensive water treatment and disposal. eMVAPEX significantly reduces water intensity, minimizing environmental impact. Based on the above findings, as shown in **Figure 1.5**, a TRL levels between 7 and 8 (limited adoption, see **Table 1.2**) can be assigned with a medium to high GHG reduction potential.

3.2.4 Steam/Flue Gas Co-Injection Field Tests

The combination of steam and flue gas (primarily oxygen, carbon dioxide, and nitrogen) presents an emerging opportunity to optimize thermal processes for heavy crude production (Pérez 2024, Hartley 2025, and Kasala 2025). **Figure 3.6** displays a simplified lateral view of the co-injection of steam and flue gas using a SAGD configuration. Other steam injection processes, such as CSS or steamflooding, could be used as well. As seen, air and natural gas are combusted to generate steam, and the combustion gases (O_2, CO_2, and N_2) are injected together downhole. This process significantly enhances the thermal efficiency of the process, reduces the SOR, increases the reservoir pressure, and lowers the overall energy requirements, making the process more economically and environmentally viable.

The production mechanisms associated with the steam/ flue gas co-injection are associated with the reduction in crude oil viscosity due to heat transfer and CO_2 solubilization in the HCO, decreased heat losses, and expansion of the heating zone

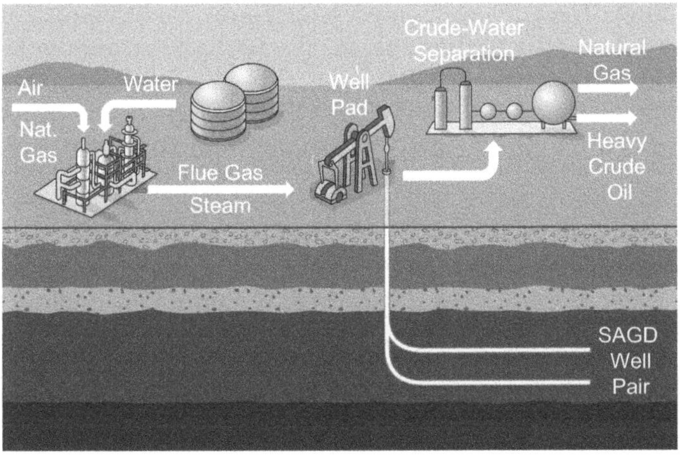

FIGURE 3.6 Simplified lateral view of the co-injection of steam and flue gas using a Steam-Assisted Gravity Drainage (SAGD) configuration.

due to the low thermal conductivity of the NCG, such as N_2, and increased reservoir pressure due to the nitrogen content (Pérez 2024, and Kasala 2025).

To address the issue of flue gas corrosion, the O_2 and CO_2 concentrations are reduced from 2% to ppm levels for the first component and from 12 to 3.5% for the second (Hartley 2025). Another concern is the portion of non-burnable CO_2 and N_2 injected that is produced back to the surface. The resulting gas, which contains CO_2 and N_2, is often blended with pipeline natural gas (see **Figure 3.6**), which lowers the heat content of the gas burned at the central processing facility (Hartley 2025)

Several field tests have been documented in the literature in USA, Canada, and China (Pérez 2024). Yee and Stroich detailed the co-injection of steam and flue gas into a mature SAGD well pair at the Dover Project, Alberta, Canada (Yee 2004). Building on the success of their previous steam and natural gas co-injection pilot, the authors conducted a four-month flue gas injection test in 2001 to explore the feasibility of using this more cost-effective gas. They reported that flue gas successfully replaced natural gas during the wind-down operation, and no signs of corrosion were detected in the steam injection and fluid production lines (Yee 2004).

Several field tests have been conducted in China, specifically in the Bohai, Xinjiang, and Shenli Fields (Zhong 2022). A modular thermal fluid generator was designed and operated for steam generation and the co-injection of steam and flue gas in both land-based and offshore oil production platforms. Several CSS examples were reported, with oil production rates two to three times higher than those of conventional cold production. The co-injection of steam and flue gas also resulted in an increase in oil recovery compared to the steam-only counterparts (Zhong 2022).

As discussed in this section, steam/flue gas co-injection technologies for heavy oil production have been successfully field-tested in a few parts of the world (TRL levels between 7 and 8, limited deployment, see **Table 1.2**). Although there have been no detailed studies, it has been reported that GHG emission reduction is 20–25%

compared to SAGD in a 2017 field project in Canada (ConocoPhillips 2017). Based on this information, the CI of steam/flue gas co-injection can be estimated at around 63 kg CO_2e/BBL, in contrast to the SAGD value of ~80 kg CO_2e/BBL (see Section 5.1 for details). These results suggest a medium potential for further reducing GHG emissions. (see **Figure 1.5**).

3.3 ELECTROMAGNETIC HEATING (RADIOFREQUENCY)

The use of RF and microwave (MW) energy for subsurface dielectric heating has shown potential for applications in thermal EOR, particularly in situations where steam injection is not feasible or optimal (Fanchi 1990, Kasevich 1997, Islam 1999, Ovalles 2002, Chhetri 2008, Bientinesi 2013, Bera 2015, Ovalles 2019, Gomaa 2024, and Rodriguez-2024). RF radiation operates at frequencies ranging from 10 to 100 MHz, while MW energy spans from 300 MHz to 300 GHz.

Before discussing the work presented in the literature, it is essential to discuss the fundamentals of electromagnetic heating and the differences between conventional resistive heating and dielectric heating techniques. As shown **in Table 3.4**, resistive heating typically operates at low frequencies (<1 kHz), whereas dielectric heating operates within the RF or MW regions. Literature indicates that resistive heating is directly proportional to the resistance of the material (Csanyi 2011). For dielectric heating, the power absorbed by a unit of volume is directly proportional to the dielectric constant (ε'), the loss tangent ($\tan \delta$), or the conductivity of the material (σ) at a given frequency (Sahni 2000). In general, the RF/MW heating of formation fluids and porous media can lead to an improved mobility of the oleic phase relative to the aqueous and gas phases, with a concomitant increase in oil production.

Based on physical and numerical simulations, as well as the limited number of field tests, several advantages of RF/MW heating technologies can be identified (Ovalles 2019). First, electromagnetic energy can be more energy efficient in deep reservoirs because formation heating can be optimized to reduce the viscosity of heavy oil. In contrast, saturated steam may far exceed optimum temperatures due to higher reservoir pressures. Additionally, RF/MW is more energy efficient in shallow

TABLE 3.4
Comparison between Resistive Heating and Dielectric Heating Techniques

Type of Heating	Resistive Heating	Dielectric Heating
Frequency	Low Frequency (< 1 kHz)	High frequency (1 kHz–10 GHz)
Equation	$P = R\,I^2$	P: $2\,\pi\,\upsilon\,\epsilon o\,\epsilon'\,\tan\delta\,E^2$, where P: $\sigma\,E^2$ $\sigma = 2\,\pi\,\upsilon\,\epsilon o\,\epsilon'\,\tan\delta$
Nomenclature (Units)	P: power (W/m^3); R: resistance (Ohm); I current traveling through the resistor (ampere [A])	P: Power (W/m^3); υ: frequency (Hz); ϵo: free space permittivity = 8.854×10^{-12} (F/m); ϵ': dielectric constant (no units); $\tan\delta$: loss tangent factor (no units); E: electric field (V/m); σ: conductivity (S/m)

wells, near thief zones (i.e., gas caps) or thin sands, which negatively affect conventional steam flood. It can be turned "off" during peak power demand to save energy and operating costs. Furthermore, electromagnetic heating could be coupled with direct solar or geothermal energy, potentially reducing the CI of the process. RF at lower frequencies provides more uniform heat distribution than steam because reservoir heterogeneity is not a barrier to RF radiation. In contrast, low-permeability formations are natural obstacles to steam, solvents, and immiscible gases. Finally, RF/MW heating can reach the very high temperatures needed for the subsurface upgrading of HCOs, does not require a water source, and avoids the swelling of clays (such as Montmorillonite and Illite), which can occur with steam condensate. These advantages position electromagnetic heating as an environmentally friendly technology, making it ideal for the sustainable production of HCO and bitumen.

Conversely, electromagnetic heating suffers from several drawbacks, including high electricity consumption, energy density around the antenna, power losses in the transmission lines, and the relatively high cost of RF/MW power supplies (Ovalles 2019).

Several laboratory experiments of RF/MW heating have been conducted in the literature. These studies have demonstrated that hydrocarbon-containing sands can absorb RF and MW to reach high temperatures (200°C–650°C) very rapidly in the absence (Cambon 1978, Depew 1991, Mutyala 2010, and Ovalles 2002) or presence of chemical additives including activated carbon, iron powder, iron oxide, NaOH, molybdic acid anhydride, and Fe-, and Cu-containing nanoparticles (Jackson 2002, Bientinesi 2013, Greff 2011, and Wheeler 2016).

Numerical simulations of the RF/MW heating process have been studied. Bogdanov *et al.* (2016) studied the production mechanism for butane injection in a typical Athabasca reservoir using both vertical and horizontal antennas. They observed significant variations in oil recoveries (~10%) depending on the mechanism of solvent mixing with the bitumen, whether by molecular diffusion or mechanical dispersion (Bogdanov 2016). In another study, Ovalles *et al.* (2020) demonstrated through numerical analysis that incorporating a low-loss zone around the central emitter significantly improved energy and temperature distribution, achieving greater penetration of around 12 m compared to cases without it.

Wang and colleagues investigated the effects of using three antenna arrays during the RF heating process, focusing on how their distance and operating parameters (voltage and frequency) influence heating patterns (Wang 2018). Their simulation results revealed that the frequency and voltage applied to each dipole antenna affected the heating pattern, including the range and direction of the heating (Wang 2018). After tuning input parameters, in a different study, Guan and coworkers predicted a temperature profile consistent with the field test observations (Guan 2021). They concluded that well-designed RF heating field pilots, along with the tuning of electromagnetic energy and reservoir parameters in simulation models, are crucial for evaluating and optimizing RF heating operations (Guan 2021). Additionally, Ramcharan and colleagues conducted a numerical study on the feasibility of oil extraction from Trinidad tar sands using radio frequency heating (Ramcharan 2016). Their results indicated an oil recovery in the 30–60% range. The energy balance analysis suggested that commercial extraction of oil from Trinidad tar sands through RF heating is viable, especially with oil prices exceeding US$25 per barrel (Ramcharan, 2016).

Based on the previous laboratory-scale tests and numerical simulations, the feasibility of using RF and MW radiations for dielectric heating of HCO-containing sands has been successfully demonstrated (Ovalles 2019). Next, the field tests reported in the literature will be examined.

3.3.1 Early Field Tests

In the 1980s and 1990s, several RF/MW field tests were conducted that showcased the potential of this technology for thermal EOR of HCOs (Bridges 1979, Sresty, 1986, and Kasevich 1994). In a pilot study in Oklahoma, Bridges and colleagues employed three waveguides inserted into the ground to heat a limited section of an oil-bearing formation (Bridges 1979). During this field test, 300 MHz radiation was used to elevate the temperature near the wellbore from 18°C to 100°C. At a distance of 15 ft from the well, the temperature reached 33°C. The authors noted an increase in the production of 6°API crude oil, although no further details were provided (Bridges 1979).

Sresty *et al.* (1986) conducted laboratory and field-scale tests to recover bitumen from tar sand deposits near Vernal (Utah) using RF heating. Their recovery methods included replacing the heated bitumen with sodium silicate solutions, gravity drainage, and autogenously generated steam and hydrocarbon gases. Two small-scale field experiments were carried out in the Asphalt Ridge tar sand deposit near Vernal. They reported that about ~150 BBL of the deposit were heated with RF energy, resulting in a recovery of 35% of the OOIP during the 3-week testing period (Sresty 1986).

Kasevich *et al.* (1994) conducted a field test in Bakersfield, California, in a 600–700 ft depth-reservoir in 1994. The authors employed a 25-kW surface generator operating at 13.56 MHz, along with a coaxial line and an antenna, to deliver energy downhole. Their results demonstrated the technical feasibility of generating, transmitting, and focusing electromagnetic energy into an oil-bearing formation, reaching temperatures of 120°C at a distance of 10 ft from the well within a period of 20 h (Kasevich 1994).

Ovalles et al. (2018) coupled the electromagnetic simulator AxREMS™ to the commercial reservoir simulator STARS™ to model RF heating using three different shaped antenna designs (straight dipole, concave, and convex designs). Their results of the static (no oil production) numerical simulation of RF dielectric heating of the three different antenna designs showed that the concave unit offers a more uniform radiation pattern and temperature profile than the straight dipole, and convex counterparts (Ovalles 2018). These improved energy and temperature distributions represented a step forward in the development of RF downhole heating technology, which was used for the reservoir simulation of RF heating.

A reservoir model was developed with seven sands (over- and under-burden and five oil-containing sands) and utilized for the dynamic simulation of downhole RF dielectric heating (Ovalles 2018). Initial temperature and pressure were 115°F and 390 psi, respectively. To account for the effect of heterogeneities, Sands 2 and 5 of the five oil-containing zones had higher permeabilities (~600 and ~800 milli Darcy) than the other three (~150 milli Darcy). A 148-ft vertical, concave-shaped dipole antenna operating at a 1.8 MHz frequency (3/4 λ) and 2 kW/m power was numerically simulated. The conditions used were as follows: three years, antenna power: 90 kW at 1.2 MHz, RF, and production wells were completed in the five central sands. One Base Case and six prediction cases were evaluated, starting the RF heating from times equal to 0, 90, and 180 days, with spacings between the RF antenna and the producer of 30and 50 ft.

Figure 3.7 shows that there is accelerated oil production for the cases in which oil production begins immediately, compared to the Base Case. In general, the RF heating process yields approximately a 2-fold increase in the cumulative oil after 700 days of heating compared to the Base Case, regardless of the moment of starting the production (**Figure 3.7**). The highest cumulative oil production occurs in the scenario where the producer is placed 50 ft updip and undergoes 180 days of RF pre-heating.

Figure 3.8 shows the temperature distributions after 690 days of RF heating in the latter case. As anticipated, the temperatures are higher around the RF antenna, reaching values in the 340°F–400°F range. Overall, the temperature distribution shown in **Figure 3.8** appears to align with the fluid movement from the RF heater to the producer. This pattern can be explained by the model permeabilities. As noted, layers 2 and 5 of the heavy oil model reservoir have higher permeabilities (approximately 600 and 800 milli Darcy) compared to the other three layers, which have permeabilities around 150 mD.

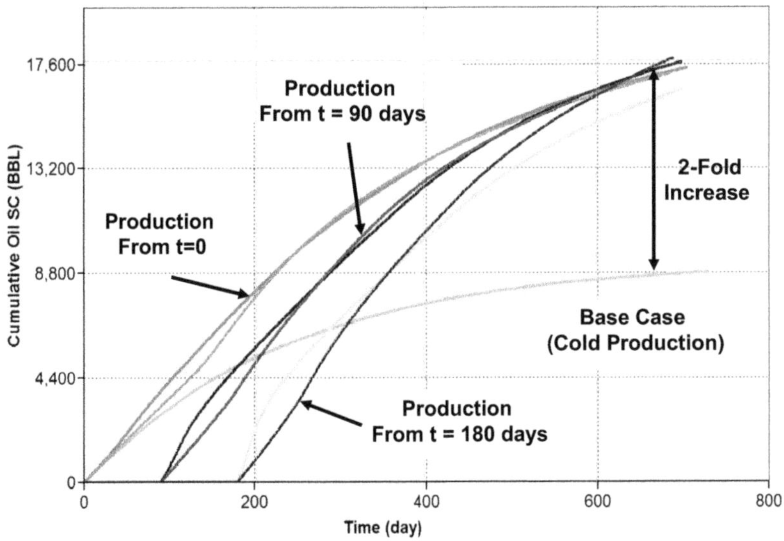

FIGURE 3.7 Simulated cumulative oil production (in BBL) for RF heating from the heavy oil reservoir model for the Base Case and the six prediction cases with spacing between the antenna and producer of 30 and 50 ft.

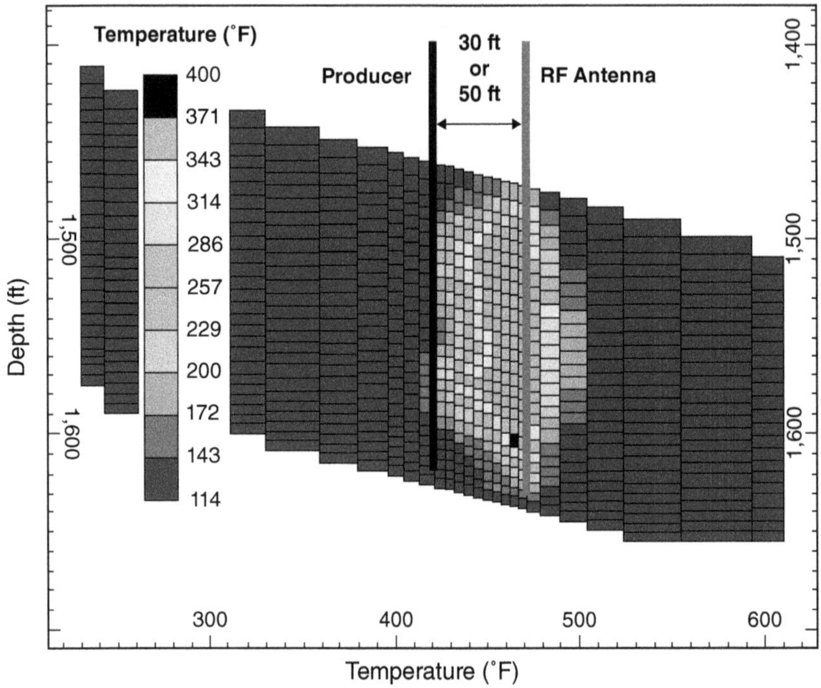

FIGURE 3.8 Simulated temperature distribution of RF heating with a 50 ft distance between the antenna and the producer, with production started after 180 days of heating.

Based on the auspicious results from the modeling studies, a concave-shaped antenna was fabricated and installed in a well (run-in hole, RIH) in the Coalinga Oil Field. Before applying power to the antenna, an integrity analysis was conducted, revealing less than 1% reflected power. This data indicated that the antenna design and the formation parameters were well-matched. However, after applying 5–10 kW of RF power to the antenna, at least two permanent shorts occurred, preventing further testing. At that point, it was decided to terminate the field test. Although the outcomes did not meet the desired objectives, they demonstrated that an antenna system could be successfully constructed and installed in the well, with properties that were compatible with the mineral formation.

3.3.2 Acceleware RF XL Field Test

Between 2018 and 2023, the Calgary-based company Acceleware conducted a field pilot at the Marwayne oilfield (Cold Lake Oil Sands region), near the Alberta–Saskatchewan border (Acceleware 2025a and 2025b). The project evaluated the potential of Acceleware's XL technology to utilize RF energy to electromagnetically heat oil sands for the EOR of HCOs and bitumen. The pilot's goals were to validate pathways to significantly lower GHG emissions, eliminate the need for fresh water in the extraction process, and noticeably lower both capital and operating costs.

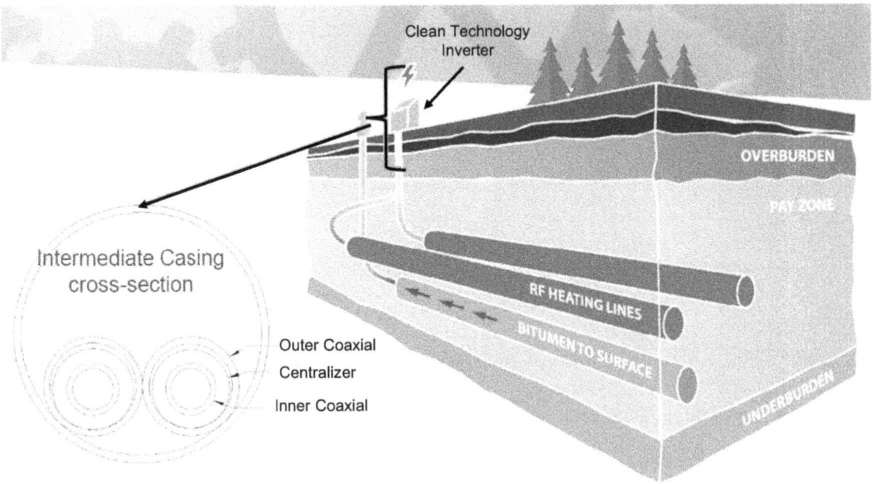

FIGURE 3.9 Schematic diagram of the Acceleware RF XL heating well.

Image courtesy of Acceleware Ltd. Calgary, Alberta, Canada.

Additionally, RF XL offers the potential to tap into vast reserves of oil previously deemed inaccessible (Acceleware 2025a).

Figure 3.9 presents a schematic diagram of the Acceleware RF XL heating well. As shown, it consists of two horizontal RF heating radiators positioned above the bitumen-producing well, all located within the oil-bearing formation, also known as the payzone. The two RF emitters are connected to the Clean Technology Inverter (CTI) on the surface via coaxial cables that run inside a casing. For the Marwayne pilot project, a three-well system (two radiators and one producer) was constructed utilizing conventional oil and gas (O&G) technologies, and the electrical integrity of the system was successfully verified. As mentioned, this process heats the oil sand electromagnetically, and the crude oil is produced through gravity drainage using the bottom horizontal producer. This production mechanism is similar to SAGD, but it does not utilize steam, making it environmentally friendly (Acceleware 2025a and 2025b).

The core component of Acceleware's RF technology is the CTI. As illustrated in **Figure 3.10**, the CTI can be powered by either grid electricity or renewable sources (wind, solar, or battery). This inverter converts electricity into RF power of up to 2 MW at frequencies ranging from 20 to 120 kHz with an efficiency of 97%. It has been successfully operated for up to seven months. The low-cost RF power generated can be used not only for EOR of HCOs but also for a variety of applications, including mining and mineral processing, concrete production, carbon capture, agri-food drying, hydrogen production, and other environmentally friendly fuels (Acceleware 2025c).

The Marwayne pilot project achieved a maximum power of >200 kW with a current of 400 A for 142 days of uninterrupted heating. The pump in the production well was shut down in October 2022. With no cooling effect of the produced fluid,

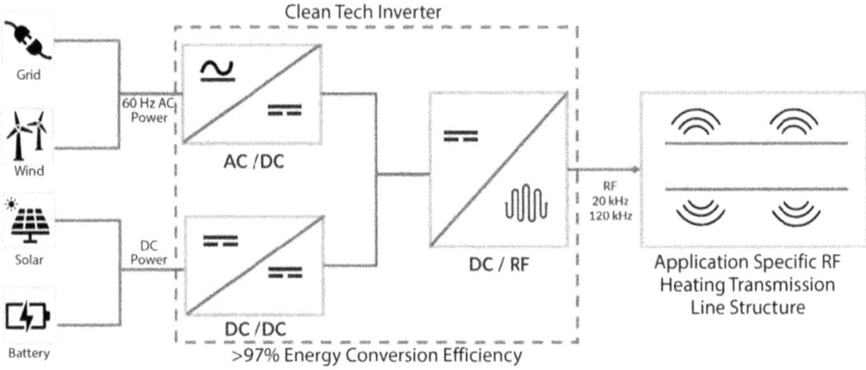

FIGURE 3.10 Diagram of Acceleware Clean Technology Inverter (CTI).

Image courtesy of Acceleware Ltd. Calgary, Alberta, Canada.

the temperature rose several degrees above the formation temperature. This finding gave evidence of RF heating in the reservoir and production well. With the success of this pilot, Acceleware is now focused on refining downhole components to improve performance and longevity, as well as further testing and validation to move closer to full-scale commercial deployment (Acceleware 2025c).

Acceleware commissioned a third-party LCA to assess the environmental benefits of the project (Okoniewski 2025). This study involved creating a detailed inventory of all the significant material, energy, and water inputs to the project and baseline scenario processes, as well as discharges of air emissions, wastes, and effluents. The baseline scenario chosen for comparison was a large-scale SAGD project. The results indicated that GHG emissions were reduced by approximately 72% in Canada and 73% in the rest of the world, while freshwater consumption for bitumen extraction in both regions would be completely eliminated (Okoniewski 2025). The estimated CI was an average of 18.8 kg CO_2e/BBL.

At the moment of writing this monograph, Acceleware is seeking funding for the RF XL 2.0 Demonstration: a commercial-scale project that builds on work performed to date and could showcase RF XL's ability to enhance recovery in heavy oil reservoirs and increase production while decarbonizing (Acceleware 2025b). On December 9, 2025, Acceleware announced a farm-in agreement for one section of land targeting the Mannville Stack in Saskatchewan (Acceleware 2025d). Under the Agreement, Acceleware plans to drill an RF XL 2.0 horizontal well pair in 2026 (Acceleware 2025d).

Based on the Marwayne field test results, it can be concluded that the Acceleware RF XL technology is at a TRL level 6 (successful field test, see **Table 1.2**). The average CI is approximately ~19 kg CO_2e/BBL, representing a ~84% and 77% reduction compared to steamflooding (Base Case) and SAGD, respectively (see Section 5.1 for details). Therefore, this technology demonstrates a medium to high potential for GHG reduction, as illustrated in **Figure 1.5**.

3.3.3 ESEIEH Field Test

Harris, Laricina, Nexen & Suncor formed a CAD$33-million consortium called the Enhanced Solvent Extraction Incorporating Electromagnetic Heating (ESEIEH™) (Rassenfoss 2012, Trautman 2014a and 2014b, Despande 2015, Wise 2016, and Morris 2021). The consortium is 50% financed by Alberta's Climate Change and Emission Management Corporation (CCEMC). In 2015, Laricina withdrew from the consortium and was replaced by Devon Canada Corp. for the later phases of the project (Bayestehparvin 2018).

The process involves the use of RF heating coupled with hydrocarbon solvent addition, with chain lengths from C2 to C5, for EOR of HCO recovery. As illustrated in **Figure 3.11**, RF emitted from the antenna in a SAGD-type configuration preheats the hydrocarbon-containing formation, allowing the solvent to dissolve the bitumen with the concomitant increase in oil production. The hydrocarbon solvent is introduced from the top horizontal well (**Figure 3.11**), and the oil is produced from the bottom. This design eliminates the need to heat the formation to steam temperatures, significantly reducing GHG emissions. The solvent decreases the viscosity of the bitumen, enhancing its mobility and thus facilitating heavy oil EOR. Additionally, this process may enable subsurface upgrading through solvent deasphalting (Trautman 2014a, Despande 2015, and Wise 2016).

Wise and Patterson conducted an economic evaluation of ESEIEH™ across three different scenarios (Wise 2016). The scenarios included a greenfield with dilbit export, a greenfield with undiluted rail export, and an ESEIEH™ pad added adjacent to an existing SAGD cogeneration facility. The results indicated that the process is economically viable at a price of US$50/BBL of West Texas Intermediate (Wise 2016 and Morris 2021).

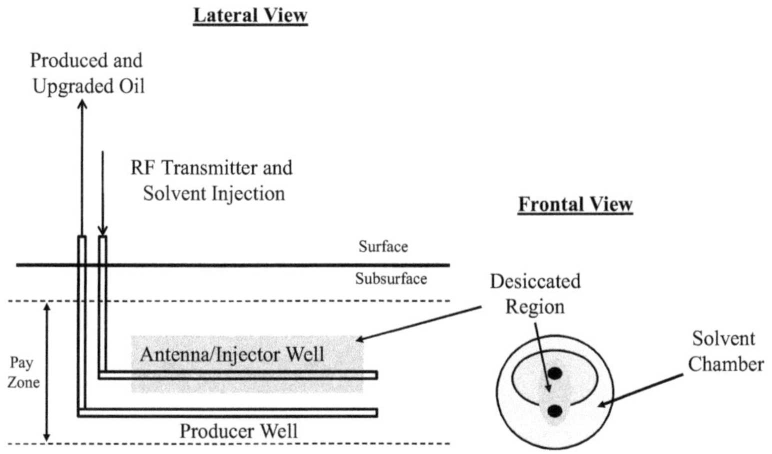

FIGURE 3.11 Lateral and frontal diagram of the Enhanced Solvent Extraction Incorporating Electromagnetic Heating process (ESEIEH™).

The ESEIEH™ Project was designed to be executed in two phases during 2011–2020. In Phase 1, a proof of concept for the RF technology was completed, along with the "Mine Face" test at Suncor's North Steepbank Mine. The results from this phase were used to verify RF energy penetration and absorption rates, revealing good temperature correlations between the coupled electromagnetic reservoir simulator and the model (Trautman 2014a). Additionally, the ESEIEH Project Team successfully executed the mine face test by developing a 10–15 m antenna able to irradiate RF up to 4 kW/m at 6.78 MHz with an average of 26 kW. Temperatures in the 35°C–130°C range were measured at 5 m and 2.5 m from the antenna, respectively (Morris 2021).

In Phase 2 of the ESEIEH™ Project, a small-scale field pilot was conducted at Suncor's Dover site in 2014. This pilot included one horizontal well pair that is 100 m long, three vertical observation wells, and all the necessary instrumentation to assess system behavior and analyze the technology's performance (Morris 2021). The results confirmed the effective penetration of electromagnetic energy into native, heterogeneous oil sands. The modeling demonstrated the potential for bitumen production, with a recovery exceeding 70% of the OOIP, as well as GHG reductions greater than 80% compared to SAGD. Unfortunately, due to the onset of the COVID-19 pandemic, challenging market conditions, and a reduction in Research and Development (R&D) spending, the ESEIEH™ Project was terminated in December 2021 (Morris 2021).

Safaei and coworkers (2019) evaluated the energy and GHG emissions' footprints of bitumen extraction using the ESEIEH™ technology using two pathways. In Pathway 1, the produced gases were captured to minimize solvent losses, while Pathway 2 involved utilizing these gases as fuel. Their findings indicated that energy consumption and GHG were 487.2 and 375.5 MJ/BBL, and 77.7 and 59.77 kg CO_2e/BBL of bitumen, respectively (Safaei 2019). In a separate study, Safaei conducted a Monte Carlo simulation to decrease uncertainty in the CI of the ESEIEH™ process (Safaei 2019). The results showed that GHG emissions varied in the range of 10–88 kg CO_2e/BBL of bitumen. The wide range is primarily due to the electricity source, as electricity accounts for 77% of the total energy required. GHG emissions are at their lowest (10 kg CO_2e/BBL) when the source of electrical energy is biomass. The highest emissions (65 CO_2e/BBL) were obtained when the current Alberta electricity grid is used. Factors such as antenna efficiency and the reservoir's cumulative electricity-to-oil ratio had the most significant impact on overall GHG emissions (Safaei 2019 and 2019).

Based on the comprehensive field test results presented in this section, it can be safely concluded that the ESEIEH™ technology is at a TRL level 6 (successful field test, see **Table 1.2**). The average CI is approximately 40 kg CO_2e/BBL, representing a ~66 and ~51% reduction compared to steamflooding (Base Case) and SAGD, respectively. Thus, a medium-high potential for GHG reduction can be assigned, as shown in **Figure 1.5**.

3.4 USE OF LOW-CARBON INTENSITY ENERGY

The use of environmentally benign energy sources for the EOR of heavy oils has been documented in the literature since the 1990s and early 2000s. Currently, this practice is not widespread due to the global availability of fossil fuels. However, with a renewed focus on environmental impacts, companies are increasingly turning to low-carbon energy sources such as wind power, sea-wave energy, solar, nuclear, and geothermal energy (Fassihi 2017, Wang 2018, Muzzafaruddin 2019, Islam 1999, Canbaz 2021, and Temizel 2022).

The utilization of wind energy has increased in the O&G industry, particularly in offshore operations and the Permian basin, where wind turbines can power platforms and oil rigs, reducing the reliance on fossil fuel-based energy sources. This transition can lead to lower operational costs, higher profitability, and reduced carbon emissions (Canbaz 2021 and Anchustegui 2024). However, wind power systems require a significant initial investment, ongoing maintenance, and may not be suitable for all geographic locations. Also, the Capital Expenditure (CAPEX) and Operating Expenditure (OPEX) costs are almost triple in offshore wind farms compared with the onshore counterparts (Canbaz 2021). Consequently, the application of wind energy for the EOR of HCO is limited.

The use of sea-wave or ocean-wave technologies as a low-CI energy source has been explored by offshore oil companies (Addinson 2018). One of the most significant advantages of sea waves is their predictability, as they are driven by stable wind patterns, particularly in offshore environments. Ocean-wave energy is environmentally friendly since it produces no chemical waste or pollution and does not adversely affect land. However, there are concerns about the potential risks to fish and marine mammals from being struck by turbine blades, particularly during high and low tides (Canbaz 2021). In April 2018, Eni deployed a wave power system in the Adriatic Sea to showcase the feasibility of wave energy technologies in O&G operations (Offshore 2018). Within six months, the buoy generated over 1 MWh of cumulative energy, operating continuously and without errors (Offshore 2019). Despite its benefits, as with wind energy, the use of sea wave technologies for the environmentally friendly extraction of HCO has not gained much popularity and will not be examined further. In the following sections, Solar Steam Generation (Section 3.4.1), Nuclear-Powered EOR (Section 3.4.2), and Geothermal Energy (Section 3.4.3) will be discussed in more detail for the exploitation of HCOs.

3.4.1 Solar Steam Generation

One of the most promising and environmentally friendly technologies for reducing costs and risks in the extraction of HCOs is the use of solar-based steam generation for EOR (Fassihi 2017, Wang 2018, Muzzafaruddin 2019, Islam 1999, Canbaz 2021, and Temizel 2022). Many heavy oilfields are situated in remote and isolated areas such as deserts, arctic zones, and offshore platforms, making the continuous supply of electricity or natural gas challenging. **Figure 3.12** shows a simplified lateral view of the solar steam generation for EOR of heavy oils using a SAGD configuration. Instead of burning natural gas to heat steam for injection, solar-EOR utilizes solar technology, including panels, mirrors, and light concentrators, to harness the sun's energy for converting water into steam. Other well configurations could be used, such as CSS and steamflooding.

Solar-based projects, like the one illustrated in **Figure 3.12**, provide numerous advantages. Once the initial investment is made, solar-based systems do not incur the

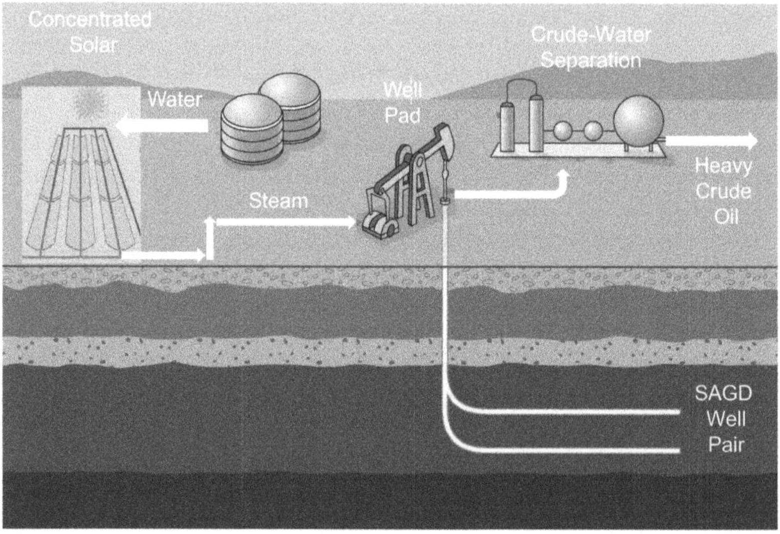

FIGURE 3.12 Simplified lateral view of the solar steam generation for EOR of heavy oils using a SAGD configuration.

high marginal costs of production as gas-based counterparts. They are not affected by market fluctuations in natural gas prices. The economics of the heavy oil project revealed that energy costs account for 45% of the total cost and more than 65% of the OPEX per barrel. Utilizing solar energy could reduce these costs by up to 40% and increase efficiency by 60–80% when thermal storage is incorporated (Al-Yatama 2018). However, thermal storage can be expensive, leading to the consideration of a hybrid system (Avaritsiotis 2018). In this approach, solar steam is injected during the day, while steam generated from natural gas is injected at night. By maintaining a constant rate of steam injection, gas consumption can be reduced by up to 25%. Also, by injecting more steam during the day and less at night, it is possible to decrease gas consumption by as much as 80% without the need for costly thermal storage (Avaritsiotis 2018).

Furthermore, solar steam technologies offer significant reductions in CO_2 emissions and carbon intensities associated with thermal oil production, while also extending the life of depleted and unprofitable oilfields and facilities (Fassihi 2017, Muzzafaruddin 2019, and Islam 1999). Conversely, the implementation of solar steam generation in the field may be constrained by its relatively large land area requirements and the variability in annual irradiance across the Earth's surface (Kovscek 2012).

In the first case study, Chevron Technology Ventures, BrightSource, and PCL partnered to build and operate a solar thermal EOR demonstration facility from 2011 to 2014 (PCL 2011 and Coalinga 2014). **Figure 3.13** depicted an aerial view of the 29 MW thermal solar-to-steam unit designed to support the EOR efforts at Chevron's Oil Field in Coalinga, California. The tower's height was reported to be 327 feet, and the project included a total of 3,822 heliostats and 7,644 mirrors (PCL 2011).

FIGURE 3.13 Aerial view of the Coalinga Solar to Steam Demonstration Plant built by BrightSource for Chevron Technology Ventures.

The pilot project consistently surpassed its annual production targets, effectively demonstrating the capability of solar thermal technology to support the EOR efforts in California in a clean and cost-effective manner. Despite its innovative approach, the project was retired in 2014, just three years after it was commissioned. The reasons for this decision likely include economic fluctuations in oil prices, operational challenges, and changing energy strategies within Chevron (PCL 2011 and Coalinga 2014).

The second case study is the Belridge Solar Project in Kern County, California, which was developed by Aera Energy in partnership with GlassPoint Solar (Dittrick 2018 and Power 2018). The Belridge Oil Field produces more than 80,000 BBL/D of heavy oil, which is recovered using steam injection technology. It is jointly owned by Shell and ExxonMobil. The Belridge Solar Project was originally designed to integrate solar thermal and photovoltaic energy to produce 12 million barrels (MBBL) of steam annually using an 850 MW solar thermal facility, and to generate 26.5 MW of electricity via photovoltaic panels (Power 2018). The solar thermal facility was planned to use GlassPoint's enclosed trough technology, which utilizes a curved solar panel setup enclosed in a greenhouse, which protects them from high winds and dust (Power 2018).

Unfortunately, the solar steam generation portion of the project was canceled in 2023. However, the solar photovoltaic panels, which were designed for on-site power generation, began operating in April 2023. Berry Petroleum Company, LLC, at the South Belridge Oil Field, is projected to generate 4,207,222 kWh of electricity during its first year, all of which will be directly consumed by the oilfield. The estimated emissions reduction from solar electricity directly supplied to the oilfield is 2,150 ton CO_2e/year. The CI reduction from the baseline is 0.23 gCO_2e/MJ (Berry 2023). Berry Petroleum Company is currently considering the installation of additional solar facilities in other California fields (Berry 2024).

In the third case study, GlassPoint and Petroleum Development Oman completed the construction on the first block of the Miraah solar plant in the Amal West Oil Field in November 2017 (Fassihi 2017, Islam 1999, and Temizel 2022). This facility utilizes enclosed trough solar technology, which consists of curved mirrors inside greenhouses that focus sunlight onto pipes filled with water. This 7-MW project generates 50 ton/d of steam at 1,450 psi (10 MPa) and 594°F (312°C), which is used for EOR of HCOs. Additionally, the solar-steam project offsets approximately 300,000 ton of CO_2 annually, significantly reducing the amount of natural gas needed for steam generation (Temizel 2018 and 2022).

From the CI point of view, life cycle assessment calculations indicated that a complete solar-based steam generation system generates 0.1 g CO_2e/MJ (0.6 kg CO_2e/BBL of HCO) over the system's lifetime vs. 29.7 g CO_2e/MJ (182 kg CO_2e/BBL) for a gas- based system (Sandler 2012 and 2014). If solar steam is utilized for one-third of the total steam generation, emissions could be reduced by 10.0 g CO_2e/MJ (61.2 kg CO_2e/BBL). This fraction corresponds to daytime solar steam generation, paired with nighttime natural gas generation, to maintain consistent steam injection rates. This value represents a 60% reduction compared to steamflooding (Base Case) (Sandler 2012 and 2014). Therefore, increasing the proportion of solar steam in place of natural gas can significantly decrease GHG emissions from HCO production (Kovscek 2012).

In summary, solar-based steam generation technologies have been successfully implemented in several oilfields around the globe; however, their widespread adoption remains limited (TRL levels between 7 and 8, limited multifield adoption, see **Table 1.2**). This is likely due to their higher CAPEX. Thus, a high potential for GHG reduction can be assigned, as shown in **Figure 1.5**.

3.4.2 Nuclear-Powered EOR

The enhanced recovery of HCOs by employing nuclear-generated steam is a low CI process that provides stable energy, independent of solar cycles, weather conditions, and other seasonal variations (Fassihi 2017). In addition to stability, nuclear power plants can operate at maximum power output for over 92% of the year (Office 2021). They require less frequent maintenance and can function for extended periods before needing refueling, typically every 1.5–2 years. Moreover, fuel costs represent a smaller portion of the overall generation expenses and have avoided large quantities of GHG emissions (International 2023). Several countries have initiated or expanded their nuclear-based energy programs, and 31 nations are committing to significantly increase their generation capacity by 2050 (Yee 2025). However, this sector faces persistent challenges, including high CAPEX, lengthy construction timelines, and ongoing public concerns about safety and nuclear waste management (Yee 2025).

The integration of nuclear energy with EOR for HCO production has been examined by several authors (Bersak 2007, Badruzzaman 2008, Finan 2010, Fassihi 2017, and Canes 2024). Traditionally, the energy to produce the steam and hot water used in HCO processes comes from natural gas. As mentioned in previous sections, the use of increasingly large amounts of natural gas for heavy oil recovery presents a number of economic and environmental problems. Finan and Kadak evaluated various nuclear reactor designs that can be used for steam generation, including the Enhanced CANDU 6, the Advanced CANDU reactor, and the pebble bed modular reactor, specifically for the production of Alberta's oil sands (Finan 2010). They reported that steam produced using nuclear energy is expected to be less expensive than steam generated from natural gas when natural gas prices are above CAD 6.50 per metric million British thermal unit (MMBtu). Additionally, nuclear energy becomes competitive with natural gas plants for electricity production when gas prices range from CAD$ 10 to CAD$ 13 per MMBtu (Finan 2010). One of the primary environmental benefits of nuclear energy in this application is its potential to reduce GHG emissions. Specifically, it could lead to a reduction of up to 3.1 million ton of CO_2 per year for each 100,000 BBL/D of bitumen production using a SAGD facility (Bersak 2007).

Maio and coworkers from Idaho National Laboratory technically evaluated the integration of a high-temperature gas-cooled reactor (HTGR) with the SAGD process for bitumen recovery (Maio 2010). The authors reported that one 600 MW HTGR is required to support the recovery of 56,000 BBL/D of bitumen. Nuclear integration eliminates approximately 56 MMSCFD of natural gas consumption by substituting nuclear heat for natural gas combustion for steam generation. Consequently, CO_2 emissions are almost eliminated from the process, although there are still CO_2 emissions from imported power (Maio 2010).

Badruzzaman and colleagues from Chevron Energy Technology Company calculated the annual CO_2 reduction per barrel for HCO and bitumen production in California and Alberta, Canada (Badruzzaman 2008). Assuming a SOR of 3.32, the authors found that utilizing nuclear-generated steam resulted in a significant reduction in CO_2 emissions compared to using natural gas as a source of energy. For California, the amount of carbon dioxide not emitted is 11.6 million ton CO_2e/year or 77.4 kg CO_2e/BBL produced. For Alberta, the calculation showed 77 million ton CO_2e/year or 70 kg CO_2e/BBL produced (Badruzzaman 2008). These values correspond to 70–90% GHG emission reduction.

In summary, various studies indicate that nuclear energy could serve as a viable, low-carbon alternative for HCO and bitumen production (Canes 2024). However, there have been limited reports on the field-scale integration of nuclear-generated steam with HCO EOR, so that a TRL level 4 can be assigned (successful modeling, see **Table 1.2**). The main barriers remain economic risk, regulatory hurdles, and public acceptance. Based on the estimated amount of CO_2 not emitted (Badruzzaman 2008), it can be anticipated that the impact on GHG emission reduction is high, as shown in **Figure 1.5**.

3.4.3 Geothermal Energy

The usage of geothermal resources in electricity production has shown a significant increase worldwide in the last twenty years (Canbaz 2021, Cano 2022, and Temizel 2022). **Figure 3.14** provides a lateral view of geothermal-based electricity production. In this process, water is pumped downhole to absorb heat from the geothermal reservoir (~3 km deep) and then brought back to the surface, where it is used to

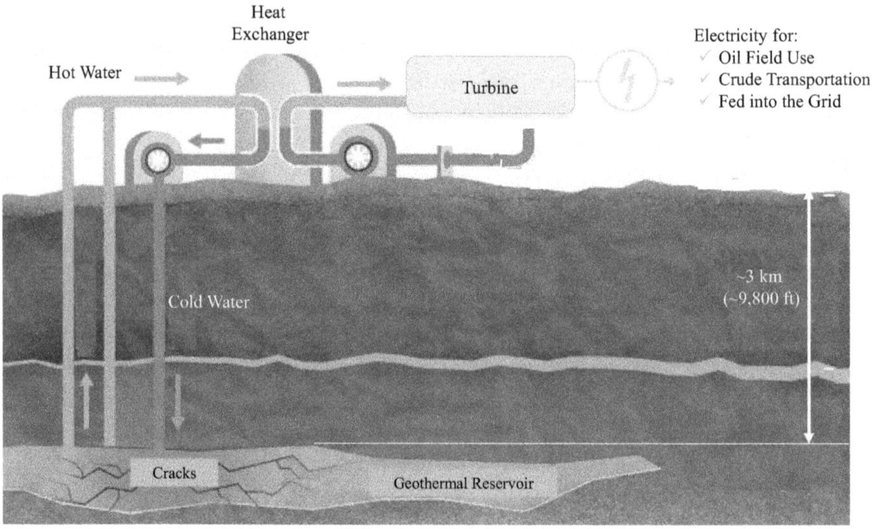

FIGURE 3.14 Lateral view of the geothermal-based electricity production.

generate electricity through a turbine. The electrical energy can be utilized to power an oilfield, for crude oil transportation, or be fed into the power grid. Also, the cooled water can be returned to the geothermal reservoir to maintain the reservoir pressure and to provide a fluid recharge. In other words, no other water source is required, and no surface water needs to be injected. Furthermore, geothermal energy is a reliable and constant source of power, unaffected by cyclical day/night or seasonal changes, unlike wind or solar energy, and is accessible throughout the year (Canbaz 2021, Cano 2022, and Temizel 2022).

Other researchers have proposed using geothermally generated hot water directly for EOR of HCOs. This hot water can also be utilized for heating greenhouses or nearby buildings (Ziabakhsh-Ganji 2018). A numerical simulation of the Moerkapelle field, located onshore in the Netherlands, demonstrated potential economic benefits (Ziabakhsh-Ganji 2018). Recently, Chevron New Energies submitted an application to Sonoma County, California, for a 20 MW geothermal power plant (GeoZone 2024). The main objective of this project is to generate electricity; however, the infrastructure and thermal output could also support nearby oilfield activities, such as steam injection for heavy oil recovery (GeoZone 2024).

The Earth's total geothermal energy is estimated to be 43 million exajoules (EJ), and it is located within 3 km of the surface (Cano 2022). Approximately 1.2 billion TWh of this energy could potentially be converted into electricity. This large amount of energy is a clear demonstration of the potential of geothermal energy for power generation (Cano 2022). Moreover, there are thousands of O&G wells drilled and abandoned due to several reasons (Canbaz 2021 and Cano 2022). Utilizing these abandoned wells for oil production from geothermal resources has gained attention. This approach could lead to considerable energy savings and help reduce drilling costs, which represent about 50% of the total expenses in a geothermal project.

Another advantage of geothermal energy is the relatively low GHG emissions. Cano and coworkers reported that, on average, conventional geothermal plants (flash or dry steam) produce 8.2 kg CO_2e/MWh^{-1} (~13.8 kg CO_2/BBL) (Cano 2022). In comparison, solar, wind, and hydropower emit lower amounts: 1.1 kg CO_2/MWh^{-1} (~1.85 kg CO_2/BBL), 0.02 kg CO_2/MWh^{-1} (~0.03 kg CO_2/BBL), and 0.8 kg CO_2/MWh^{-1} (~1.34 kg CO_2/BBL), respectively (Yousefi 2019, Cano 2022, and Low-Carbon 2024). This means that geothermal energy has one of the lowest carbon intensities among renewable energy sources, especially when compared to fossil fuels and other renewable options.

However, the development of geothermal energy is constrained by several factors, including the risks associated with subsurface exploration and the high initial capital costs required for drilling and constructing facilities (Canbaz 2021, Cano 2022, and Temizel 2022). Furthermore, many sites with potential for geothermal energy are located at significant distances from populated areas and major HCO production sites (Energy 2022).

Based on the limited field tests disclosed in the literature, it can be assessed that the use of geothermal energy to generate steam for EOR of HCO is at TRL level 5 (initial field tests conducted under favorable conditions, see **Table 1.2**). Its progression to higher TRLs is likely hindered by the relatively high CAPEX and exploration risk. Based on relatively low GHG emissions (Cano 2022), it can be anticipated that the impact on GHG emission reduction is high, as shown in **Figure 1.5**.

3.5 CONCLUSION

- Thermal methods are the most commercially utilized EOR methods for HCO production and include hot water injection, CSI, steamflooding, and SAGD (TRL levels between 7 and 9). Their main challenges include high energy and water consumption, the need for expensive water treatment to generate steam, substantial heat losses, and significant GHG emissions.
- Several emerging, environmentally friendly technological approaches have also been developed (TRL levels between 7 and 9), which reduce GHG emissions while maintaining or improving oil recovery factors by increasing recovery up to **30–70%** of the OOIP. Among the advanced processes with high potential for reducing GHG emissions, we can find steam/solvent co-injection (SAGD/Solvent and CSS/solvent), NCGs (Steam/NCG, eMVA-PEX, and eMSAGP), electromagnetic heating, and solar steam generation.
- Low CI energy for the production of HCOs, such as solar steam generation, nuclear-powered EOR, and geothermal energy, represents alternative routes with high potential for reducing GHG emissions.

REFERENCES

"Acceleware Announces RF XL 2.0," 2025a, June, https://www.acceleware.com/post/acceleware-announces-rf-xl-2-0. Retrieved July 17, 2025.

Acceleware Innovations, 2025b "Electromagnetic Heating Technology Using Radio Waves to Heat and Mobilize Heavy Oil and Bitumen" April 22, Electromagnetic Heating Technology. Retrieved April 30, 2025.

"Acceleware Ltd. Enters into Farm-In Agreement in Saskatchewan Mannville" 2025c, December, Acceleware Ltd. Enters into Farm-In. Retrieved December 11, 2025.

"Acceleware Ltd. Reports Fourth Quarter-2024 Financial and Operating Results," 2025d, April, https://www.acceleware.com/post/acceleware-ltd-reports-fourth-quarter-2024-financial-and-operating-results. Retrieved April 30, 2025.

Addinson, V., 2018, "Offshore Oil Companies Dip Into Ocean Wave Energy To Power Up" Hart Energy, September 9. Retrieved August 5, 2025.

Ahmadi, M., 2023, *Sustainable In-Situ Heavy Oil and Bitumen Recovery*. Gulf Professional Publishing, Cambridge, MA, USA.

Alberta Energy Regulator, 2014, "Devon Subsurface Performance Presentation," October 2014, p 84. Retrieved June 12, 2017.

Alberta Energy Regulator, 2015, "Annual Surmont SAGD Performance Review," http://www.aer.ca/documents/oilsands/insitu-presentations/2015AthabascaConocoSurmontES-SAGD11596.pdf. Downloaded on June 11, 2017.

Alberta Energy Regulator, 2016a, "Connacher Performance Presentation," p. 35. https://www.aer.ca/documents/oilsands/insitu-presentations/2016ConnacherGreatDivide10587.pdf. Downloaded on June 12, 2017.

Alberta Energy Regulator, 2016b, Statoil Annual Performance Presentation, March 2016, p. 57. https://www.aer.ca/documents/oilsands/insitu-presentations/2016AthabascaStatoilLeismer SAGD10935-Presentations.zip. Downloaded on June 12, 2017.

Alberta Energy Regulator, 2018, "Control Effectiveness Assessment of Solvent Injection Processes," December, https://static.aer.ca/prd/documents/reports/SolventInjection_ControlAssessment.pdf. Accessed May 27, 2025.

Ali, B., Kumar, A., 2017, "Development of Life Cycle Water Footprints for Oil Sands-Based Transportation Fuel Production", *Energy*, 131, 41–49, https://doi.org/10.1016/j.energy.2017.05.021

Alsabagh, A.M., Salem, A.M., Abdel-Aleim, G.M., Khalil, S.A., Roushdi, M., 2023, "Review of Recent Developments in Enhanced Oil Recovery: The Integration of Steam Injection with Chemical Additives and their Effect on Heavy Oil Recovery," *J. Pet. Mining Eng.*, 25(1), 19–34.

Alvarez, J., Han, S., 2013, "Current Overview of Cyclic Steam Injection Process", *J. Pet. Sci. Res.*, 2(3), July.

Al-Yatama, M., Al-Khulaifi, A., Gomez, F.J., Ghouti, R., Afifi, S., 2018, "Economic Benefits of Implementing Alternative Energy: A Heavy Oil Fields Case Study." Paper presented at the *SPE International Heavy Oil Conference and Exhibition*, Kuwait City, Kuwait, December. https://doi.org/10.2118/193717-MS

Anchustegui, I.H., Tscherning, R., 2024, "Offshore Oil and Gas Infrastructure Electrification And Offshore Wind: A Legal Exploration," *J. World Ener. Law Bus.*, 17(1), 35–53, https://doi.org/10.1093/jwelb/jwad035

Ardali, M., Barrufet, M., Mamora, D.D., Qiu, F., 2012, "A Critical Review of Hybrid Steam/Solvent Processes for the Recovery of Heavy Oil and Bitumen", SPE No. 159257, presented at the *SPE Annual Technical Conference and Exhibition*, San Antonio, TX, 8–10 October, and references therein. https://doi.org/10.2118/159257-MS

Austin-Adigio, M., Gates, I., 2019, "Non-Condensable Gas Co-Injection with Steam for Oil Sands Recovery," *Energy*, 179, 736–746. https://doi.org/10.1016/j.energy.2019.05.034

Avaritsiotis, J., 2018, "Solar Energy Enhanced Oil Recovery: A Case Study for Kuwait," Paper presented at the *SPE International Heavy Oil Conference and Exhibition*, Kuwait City, Kuwait, December. https://doi.org/10.2118/193801-MS

Ayodele, O.R., Nasr, T.N., Beaulieu, G., Heck, G., 2009, "Laboratory Experimental Testing and Development of an Efficient Low Pressure ES-SAGD Process", *J. Can. Pet. Soc.*, 48(9), September 54. https://doi.org/10.2118/09-09-54

Badruzzaman, A., Hedges, J., Demayo, T., Sigworth, H., 2008, "Nuclear Energy for Unconventional Fossil-Fuel Resource Recovery." Paper presented at the *SPE Annual Technical Conference and Exhibition*, Denver, Colorado, USA, September. https://doi.org/10.2118/116294-MS

Bayestehparvin, B., Farouq Ali, S.M., Abedi, J., 2018, "Solvent-Based and Solvent-Assisted Recovery Processes: State of the Art", SPE No. 179829, *SPE Res. Eval. Eng.*, August, 1–21, and references therein. https://doi.org/10.2118/179829-PA

Bera, A., Babadagli, T., 2015 "Status of Electromagnetic Heating for Enhanced Heavy Oil/Bitumen Recovery and Future Prospects: A Review", *Applied Energy*, 151, 206–226. https://doi.org/10.1016/j.apenergy.2015.04.031

Berry Petroleum Company, 2023, "Innovative Crude Oil Production Method Application", Application to California Air Resources Board for Low Carbon Fuel Standard, September 5, https://ww2.arb.ca.gov/sites/default/files/classic/fuels/lcfs/crude-oil/innovative-crude/summary_bry.pdf. Retrieved August 6, 2025.

Berry Petroleum Company, 2024, "Sustainable Business Report", 2024_04_10-Berry-Corporation_Sustainability-Report_April-2024-vF1.pdf. Retrieved August 6, 2025.

Bersak, A.F., Kadak, A.C., 2007, *Integration of Nuclear Energy with Oil Sands Projects For Reduced Greenhouse Gas Emissions and Natural Gas Consumption.* Department of Nuclear Science and Engineering, Massachusetts Institute of Technology, June. https://www.ourenergypolicy.org/wp-content/uploads/2012/03/MITWhitePaper.pdf. Retrieved August 2, 2025

Bientinesi, M., Petarca, L., Cerutti, A., Bandinelli, M., De Simoni, M., Manotti, M., Maddinelli, G., 2013, "A Radiofrequency/Microwave Heating Method for Thermal Heavy Oil Recovery Based on a Novel Tight-shell Conceptual Design", *J. Pet. Sci. Eng.*, 107, 18–30. https://doi.org/10.1016/j.petrol.2013.02.014

Bogdanov, I., Cambon, S., Mujica, M., Brisset, A., 2016, "Heavy Oil Recovery via Combination of Radio-Frequency Heating with Solvent Injection", SPE No. 180709, presented at *SPE Canada Heavy Oil Technical Conference*, 7–9 June, Calgary, Alberta, Canada. https://doi.org/10.2118/180709-MS

Brandt, A.R., 2012, "Variability and Uncertainty in Life Cycle Assessment Models for Greenhouse Gas Emissions from Canadian Oil Sands Production," *Env. Sci. Tech*, 46(2), 1253–1262, https://doi.org/10.1021/es202312p

Brandt, A.R., Unnasch, S., 2010, "Energy Intensity and Greenhouse Gas Emissions from Thermal Enhanced Oil Recovery," *Energy Fuels*, 24(August), 4581–4589. https://doi.org/10.1021/ef100410f

Bridges E.A., 1979, "Method for In-situ Heat Processing of Hydrocarbonaceus Formations", US Patent 4,140,180

Butler, R., 1999, "The Steam and Gas Push (SAGP)." *J Can Pet Technol.*, 38, No pagination specified. https://doi.org/10.2118/99-03-05

Butler, R., Yee, C.T., 2022, "Progress in the In Situ Recovery of Heavy Oils and Bitumen", *J. Can. Pet. Technol.* 41: No Pagination Specified. https://doi.org/10.2118/02-01-02

California Air Resources Board, 2023, "Calculation of 2023 Crude Average Carbon Intensity Values". Retrieved November 10, 2025

Cambon, J.L., Kyvana, D., Chavarie, C., Bosisio, R.G., 1978, "Traitement du Sable Bitumineux Par Micro-Ondes", *Can. J. Chem. Eng.*, 56, 735–742. https://doi.org/10.1002/cjce.5450560614

Canbaz, C.H., Temizel, C., Palabiyik, Y., Kor, K., Hendrandingrat, L., Bahar Hosgor, F., Aksahan, F., 2021 "Environmentally-Friendly Production and Recovery Processes for Heavy Presented at *SPE Trinidad and Tobago Section Energy Resources Conference Held Virtually* 28–30 June 2021. https://doi.org/10.2118/200897-MS

Canes, M., 2024, "Integration of Nuclear Energy with Oil Sands Projects for Reduced Greenhouse Gas Emissions and Natural Gas Consumption", September 19. Retrieved August 2, 2025

Cano, N.A., Céspedes, A., Redondo, J., Foo, G., Jaramillo, D., Martinez, D., Gutiérrez, M., Pataquiba, J., Rojas, J., Cortés, F.B., Franco, C.A., 2022, "Power from Geothermal Resources as a Co-product of the Oil and Gas Industry: A Review," *ACS Omega*, 7, 40603–40624. https://doi.org/10.1021/acsomega.2c04374

Castellanos-Diaz, O., Verlaan, M.L., Hedden, R., 2016. "Solvent Enhanced Steam Drive: Results from the First Field Pilot in Canada", SPE No. 179815, presented at the *SPE EOR Conf. Oil& Gas west Asia*, Muscat, Oman, 21–23 March. https://doi.org/10.2118/179815-MS

Cenovus Energy Inc, 2024, "Foster Creek In-situ Progress Report", Report Scheme 8623, June 30th, https://static.aer.ca/prd/documents/oilsands/insitu-presentations/2024-cenovus-foster-8623.pdf. Retrieved May 27, 2025.

Chen, Y., Dadashi, Z., Sleep, S., Brandt, A.R., MacLean, H.L., Bergerson, J.A., 2019, "Life Cycle Analysis (LCA) of GHG Emissions of Oil Sands Reference Facilities and Emerging Technologies" Final Report Phase 1, Submitted To: Alberta Innovates, Emissions Reduction Alberta. December 10th. https://albertainnovates.ca/wp-content/uploads/2021/01/LCA-Oil-Sands-Final-Report-December-10th-2019.pdf. Retrieved June 26, 2025.

Chhetri, A.B., Islam, M.R., 2008, "A Critical Review of Electromagnetic Heating for Enhanced Oil Recovery," *Pet. Sci. Techn.*, 26(14), 1619–1631. https://doi.org/10.1080/10916460701287607

Clean Surfactant Energy (CSE) Project, 2023, "Cyclic Steam Replacement," December. Accessed May 19, 2025

Coalinga, 2014, "Chevron/BrightSource Solar-to-Steam Demonstration Facility", Coalinga Enhanced Oil Recovery Project. Retrieved August 5, 2025.

Connacher Performance Report for the 2023, Calendar Year, 2023, https://static.aer.ca/prd/documents/oilsands/insitu-presentations/2024-connacher-greatdivide-10587.pdf. Retrieved May 27, 2025.

ConocoPhillips, 2017, "Catching Thieves in the Oil Sands", November 2. Retrieved June 27, 2025

Csanyi, E., 2011, Resistive heating explained in detail. https://electrical-engineering-portal.com/resistive-heating-explained-in-details. Tetrieved July 24, 2025.

Depew, M.C., Lem, S., Wan, J.K.S., 1991, "Microwave Induced Catalytic Decomposition of Some Alberta Oil Sands and Bitumens," *Res Chem Intermed*, 16, 213–223. https://doi.org/10.1163/156856791X00273

Despande, R., Wright, B.N., Watt, A. 2015, Techniques for Installing Effective Solvent Extraction Incorporating Electromagnetic Heating ("ESEIEH") Completions. Presented at the *World Heavy Oil Conference*, Edmonton, Alberta, Canada, Paper No. WHOC15-317.

Dittrick, P., 2018 "Aera Energy to Use Solar Energy for Steam Floods in Belridge Oil Field", *Oil & Gas J.*, January 1, Aera Energy to use solar energy. Retrieved August 5, 2025

Doan, L.T., Harschnitz, B., Shiga, N., Pennacchioli, E., Park, B., 2014, "NCG Co-Injection at Hangingstone Demonstration Project: Case Study and Analysis." Paper presented at the *SPE Heavy Oil Conference-Canada*, Calgary, Alberta, Canada, June. https://doi.org/10.2118/170068-MS

Energy Information Administration, 2022 "Geothermal Explained: Where Geothermal Energy is Found" February 15, Where geothermal energy is found. Retrieved August 12, 2005

Escobar, M.A., Valera, C.A., Perez, R.E., 1997, "A Large Heavy Oil Reservoir in Lake Maracaibo Basin: Cyclic Steam Injection Experiences," Paper presented at the *International Thermal Operations and Heavy Oil Symposium*, Bakersfield, California, February. https://doi.org/10.2118/37551-MS

Fanchi, J.R., Feasibility of Reservoir Heating by Electromagnetic Irradiation." Paper presented at the *SPE Annual Technical Conference and Exhibition*, New Orleans, Louisiana, September 1990. https://doi.org/10.2118/20483-MS

Fassihi, M.R., Kovscek, A.R., 2017, "Low-Energy Processes for Unconventional Oil Recovery", Society of Petroleum Engineers, ISBN (electronic): 978-1-61399-891-5. https://doi.org/10.2118/9781613994757

Fayazi, A., Kantzas, A., 2019, "A Review on Steam-Solvent Processes for Enhanced Heavy Oil/Bitumen Recovery." *Reviews Chem. Eng.*, 35(3), 393–419. https://doi.org/10.1515/revce-2017-0008

Finan, A., Kadak, A.C., 2010, "Integration of Nuclear Energy Into Oil Sands Projects", *J. Eng. Gas Turbines and Power*, 132, April. 042902-1-8. https://doi.org/10.1115/1.3098421

Friedmann, F., Smith, M.E., Guice, W.R., Gump, J.M., Nelson, D.G., 1994, "Steam-Foam Mechanistic Field Trial in the Midway-Sunset Field," *SPE Res Eng.*, 9, 297–304. https://doi.org/10.2118/21780-PA

Gauglitz, P.A., Smith, M.E., Holtzclaw, M.I., Duran, H., 1993, "Field Optimization of Steam/Foam for Profile Control: Midway-Sunset 26C," Paper presented at the *SPE International Thermal Operations Symposium*, Bakersfield, California, February. https://doi.org/10.2118/25781-MS

GeoZone (Chevron New Energies) geothermal power plant, 2024, https://www.gem.wiki/GeoZone_%28Chevron_New_Energies%29_geothermal_power_plant. Retrieved August 12, 2025

Gomaa, S., Salem, K.G., El-Hoshoudy, A.N., 2024, "Enhanced Heavy and Extra Heavy Oil Recovery: Current Status and New Trends," *Petroleum*, 10(3), 399–410. https://doi.org/10.1016/j.petlm.2023.10.001

Gordon, D., Brandt, A., Bergerson, J., Koomey, J., 2015, Know Your Oil: Creating a Global Oil-Climate Index Carnegie Endowment for International Peace. Accessed May 18, 2025

Greff, J., Babadagli, T. 2011. "Catalytic Effects of Nano-Size Metal Ions in Breaking Asphaltene Molecules during Thermal Recovery of Heavy-Oil", SPE No. 146604 presented at *SPE Annual Technical Conference and Exhibition*, 30 October–2 November, Denver, Colorado, USA. https://doi.org/10.2118/146604-MS

Guan, X., Li, G., Wang, H., Shang, S., Tokar, T., McVey, K., Ovalles, C., Wu, D., Chen, J., 2021, "Radio Frequency Heating Simulation Using A Reservoir Simulator Coupled with Electromagnetic Solver for Soil Remediation," Paper presented at the *SPE Reservoir Simulation Conference*, On-Demand, October. https://doi.org/10.2118/204003-MS

Guo, K., Li, H., Yu, Z. 2016 "In-Situ Heavy and Extra-Heavy Oil Recovery: A Review", *Fuel*, 185, 886–902, and references therein. https://doi.org/10.1016/j.fuel.2016.08.047

Gupta, S., Gittins, S., Picherack, P., 2005,. "Field Implementation of Solvent Aided Process", *J Can. Pet. Tech.*, 44(11), 8–13. https://doi.org/10.2118/05-11-TN1

Gupta, S.C., Gittins, S.D., 2006, "Christina Lake Solvent Aided Process Pilot", *J Can. Pet. Tech.*, 45(9), 15–18. https://doi.org/10.2118/06-09-TN

Hartley, T., 2025, "Reducing Costs in SAGD: NCG Co-Injection with GERI's DCSG Technology" GERI (General Energy Recovery Inc.) Publication. https://choa.ab.ca/choa-journal/. Retrieved June 11, 2025.

Hascakir, B., 2016, "How to Select the Right Solvent for Solvent-Aided Steam Injection Processes," *J. Pet. Sci. Eng.*, 146, 746–751. https://doi.org/10.1016/j.petrol.2016.07.038

Hedden, R., Verlaan, M., Lastovka, V., 2014, "Solvent Enhanced Steam Drive" SPE No. 169070 presented at *SPE Improved Oil recovery Symposium*, Tulsa OK, 12–16 April. https://doi.org/10.2118/169070-MS

Hoque, M., 2023, "Techno-Economic Analysis and GHG Emissions' Footprints of Solvent-steam-based Bitumen Extraction", A thesis submitted in partial fulfillment of the requirements for the degree of Master of Science, Department of Mechanical Engineering, University of Alberta.

Imperial Cold Lake Operations, 2023, "Directive 054 Annual Submission 2023." June, https://static.aer.ca/prd/2024-10/2024-imperial-coldlake-8558.pdf. Accessed May 27, 2025

International Energy Agency, 2023, "Nuclear Power", July 11, Nuclear – IEA. Retrieved August 11, 2025.

Islam, M. R., 1999, "Emerging Technologies in Enhanced Oil Recovery," *Energy Sources*, 21 (1–2), 97–111. https://doi.org/10.1080/00908319950014984

Ivory, J., Frauenfeld, T., Jossy, C., 2010, "Thermal Solvent Reflux and Thermal Solvent Hybrid Experiments," *J Can Pet Technol*, 49, 23–31. https://doi.org/10.2118/133202-PA

Jackson, C. 2002 "Upgrading a Heavy Oil Using Variable Frequency Microwave Energy", SPE No. 78982 presented at *SPE International Thermal Operations and Heavy Oil Symposium and International Horizontal Well Technology Conference*, 4–7 November, Calgary, Alberta, Canada. https://doi.org/10.2118/78982-MS

Jaimes D.M., Gates, I.D., Clarke, M., 2019, "Reducing the Energy and Steam Consumption of SAGD Through Cyclic Solvent Co-Injection," *Energies*, 12, 3860–3888. https://doi.org/10.3390/en12203860

Jha, R.K., Kumar, M., Benson, I., Hanzlik, E., 2012, "New Insights into Steam-Solvent Co-injection Process Mechanism", SPE No. 159277, presented at the *SPE Annual Technical Conf. and Exhibition* held in San Antonio, Texas, USA, 8–10 October. https://doi.org/10.2118/159277-MS

Kasala, E.E., Wang, J., Majid, A., Nadege, M.N., Nyakilla, E.E., 2025 "Advancement Perspectives in Enhancing Heavy Oil Recovery Using N_2- and CO_2-Assisted Steam Injection: Exploring Molecular Interactions to Field Applications," *Energy Fuels*, 39, 10200−10244. https://doi.org/10.1021/acs.energyfuels.5c01287

Kasevich, R.S., Price, S.L., Albertson, A., 1997, "Numerical Modeling of Radio Frequency Heating Process for Enhanced Oil Production." Paper presented at the *SPE Western Regional Meeting*, Long Beach, California, June. https://doi.org/10.2118/38311-MS

Kasevich, R.S., Price, S.L., Faust, D.L., Fontaine, M.F., 1994, "Pilot Testing of a Radio Frequency System for Enhanced Oil Recovery from Diatomaceous Earth", SPE No. 28619, presented at *69th Annual Technical Conference*. New Orleans, LA, USA, September 25–28. https://doi.org/10.2118/28619-MS

Kim, S., Shin, H., Park, C., Chen, Z., Lee, K., 2025, "A Review of Design Factors in Steam and Gas Push for Eco-Friendly Oil Sands Production and Its Field Application in Canada, "*J Petrol Explor Prod Tech.*, 15(8), 1–15. https://doi.org/10.1007/s13202-024-01903-7

Kovscek, A.R., 2012, "Emerging challenges and potential futures for thermally enhanced oil recovery," *J. Pet. Sci. Eng.*, 98, 130–143. https://doi.org/10.1016/j.petrol.2012.08.004

Leaute, R.P., 2002, "Liquid Addition to Steam for Enhancing Recovery (LASER) of Bitumen with CSS: Evolution of Technology from Research Concept to a Field Pilot at Cold Lake", CIM/CHOA 79011, presented at *The Thermal Management Systems Symposium*, held in Calgary Alberta, Canada 4–7 November. https://doi.org/10.2118/79011-MS

Leaute, R.P., Carey, B., 2007,"Liquid Addition to Steam for Enhancing Recovery (LASER) of Bitumen with CSS: Results from the First Pilot Cycle," *J. Can. Pet. Tech.*, 46(9), 22–30. https://doi.org/10.2118/07-09-01

Lee, J., Babadagli, T. 2021, "Mitigating Greenhouse Gas Intensity Through New Generation Techniques During Heavy Oil Recovery," *J Clean Prod.*, 286, 124980. https://doi.org/10.1016/j.jclepro.2020.124980

Li, S., Li, Z., Li, B., 2014, "Experimental study and application of tannin foam for profile modification in cyclic steam stimulated well," *J. Pet. Sci. Eng.*, 118, 88–98, https://doi.org/10.1016/j.petrol.2014.04.009

Lin, L., Ma, H., Zeng, F., Gu, Y., 2014, "A Critical Review of the Solvent-Based Heavy Oil Recovery Methods". Paper presented at the SPE Heavy Oil Conference-Canada, Calgary, Alberta, Canada, June 2014. https://doi.org/10.2118/170098-MS

Low-Carbon Power, 2024, "0% of Global Electricity is Generated from Geothermal", Geothermal Energy: Global Electricity Generation 2024 | Low-Carbon Power. Retrieved August 12, 2025.

Maio, Y. C., Gribik, A. M., Patterson, M. W., Boardman, R. D., 2010, "Nuclear-Integrated Oil Sands Recovery via Steam-Assisted Gravity Drainage" Technical Evaluation Study, Project No. 23843. Idaho National Laboratory. Nuclear-Integrated Oil Sands. Retrieved December 10, 2025.

MEG Energy Corp, 2020, "eMVAPEX Commercial Demonstration. Non-Confidential Final Report," Project ID F0160215, Prepared for: Emissions Reduction Alberta. February. eMVAPEX Commercial Demonstration. Retrieved June 12, 2025

MEG Energy Corp, 2022, "Christina Lake Regional Project 2021 Directive 54 Performance Report" Alberta Energy Regulator Commercial Scheme Approval No. 10773, June. https://static.aer.ca/prd/documents/oilsands/insitu-presentations/2022-Athabasca-MEG-Christina-Lake-10773.pdf, Retrieved June 26, 2025.

MEG Energy Corp, 2023, "Christina Lake Regional Project" 2022 Directive 54 Performance Report Experimental Scheme Approval No. 12528, June. Retrieved May 27, 2025.

MEG Energy Corp. 2024, "Enhanced Modified VAPour EXtraction R&D Operation", December 23, https://natural-resources.canada.ca/funding-partnerships/enhanced-modified-vapour-extraction-rd-operation. Retrieved June 11, 2025

Mendez, Z., Alvarez, J.M., Escobar, E., Colonomos, P., Campos, E., 1992, "Cyclic Steam Injection with Additives: Laboratory and Field Test Results of Steam/Foam and Steam/Solvent Processes." Paper presented at the *SPE Annual Technical Conference and Exhibition*, Washington, D.C., October. https://doi.org/10.2118/24632-MS

Morris, P., 2021 "Effective Solvent Extraction Incorporating Electromagnetic Heating (ESEIEH™)" Project Fibal Report to the Emissions Reduction Alberta, Submission date 2022. Retrieved July 28, 2021.

Mutyala, S., Fairbridge, C., Paré, J.R.J., Bélanger, J.M.R., Ng, S., Hawkins, R., 2010, "Microwave Applications to Oil Sands and Petroleum: A Review," *Fuel Proc. Tech.*, 91(2), 127–135. https://doi.org/10.1016/j.fuproc.2009.09.009

Muzzafaruddin, M.D.K., 2019, "Enhanced Oil Recovery," *Int. J. Pet. Petro. Eng.*, 5(4), 10–13. https://doi.org/10.20431/2454-7980.0504002

Nasr, T. N., Beaulieu, G., Golbeck, H., Heck, G., 2009, "Novel Expanding Solvent-SAGD Process ES-SAGD," *J. Can. Pet. Tech.*, 42(1), 13–16. https://doi.org/10.2118/03-01-TN

Nasr, T. N., Isaacs, E. E., 2001, "Process for Enhancing Hydrocarbon Mobility Using a Steam Additive", US Patent 6,230,814. US6230814B1 - Process for enhancing hydrocarbon mobility using a steam additive.

NIPER, 1986, Enhanced Oil Recovery Information booklet as part of the U.S. Department of Energy (DOE) EOR research program. Downloaded on March 29th, 2017 from the link https://www.netl.doe.gov/research/oil-and-gas/enhanced-oil-recovery/eor-process-drawings

Odier, C., Kerdraon, M., Lacombe, E., Delamaide, E., 2021, "Very High Temperatures Steam Foam Additives." Paper presented at the *Abu Dhabi International Petroleum Exhibition & Conference*, Abu Dhabi, UAE, November. https://doi.org/10.2118/207464-MS

Office of Nuclear Energy, 2021, "Nuclear Power is the Most Reliable Energy Source and It's Not Even Close", Department of Energy. Retrieved August 11, 2025.

Offshore Energy Today, 2019, "OPT, Eni in wave energy device milestone", May 15. Retrieved August 5, 2025.

Offshore Staff, 2018, "Eni deploys wave power system in the Adriatic Sea", December 21. Retrieved August 5, 2025.

Okoniewski, M., 2025, Acceleware Chief Scientific Officer & Co-Founder, personal communications, September 16.

Orellana, A., Laurenzi, I.J., MacLean, H. L., Bergerson, J. A., 2018 "Statistically Enhanced Model of In Situ Oil Sands Extraction Operations: An Evaluation of Variability in Greenhouse Gas Emissions," *Env. Sci. & Tech.* 52 (3), 947–954, https://doi.org/10.1021/acs.est.7b04498

Orr, B. 2009, "ES-SAGD; Past, Present, and Future. Paper" SPE 129518 presented at *SPE Annual Tech. Conf. Exhib.* New Orleans, Louisiana, 4–7 October. https://doi.org/10.2118/129518-STU

Osma, L., García, L., Pérez, R., Barbosa, C., Botett, J., Sandoval, J., Manrique, E., 2019, "Benefit–Cost and Energy Efficiency Index to Support the Screening of Hybrid Cyclic Steam Stimulation Methods," *Energies*, 12, 4631. https://doi.org/10.3390/en12244631

Ovalles C., 2019, *Subsurface Upgrading of Heavy Crude Oils and Bitumen*. CRC Press, Boca Raton: Taylor & Francis Group, CRC Press; Chapter 3. https://doi.org/10.1201/9781315181028

Ovalles, C., Fonseca, A., Lara, A., Alvarado, V., Urrecheaga, K., Ranson, A., Mendoza, H., 2002, "Opportunities of Downhole Dielectric Heating in Venezuela: Three Case Studies Involving Medium, Heavy and Extra-Heavy Crude Oil Reservoirs." Paper presented at the *SPE International Thermal Operations and Heavy Oil Symposium and International Horizontal Well Technology Conference*, Calgary, Alberta, Canada, November. https://doi.org/10.2118/78980-MS

Ovalles, C., Rico, A., Perez-Perez A., Hernandez, M., Guzman, N., Anselmi, L., Manrique, E., 2002, "Physical and Numerical Simulations of Steamflooding in a Medium Crude Oil Reservoir, Lake Maracaibo, Venezuela", paper SPE-75131-MS presented at *the SPE/DOE Improved Oil Recovery Symposium held in Tulsa*, Oklahoma, 13–17 April. https://doi.org/10.2118/75131-MS

Ovalles, C., Vaca, P., Dieckmann, G. H., Dunlavey, J., Behrens, R., Dillenbeck, L., Okoniewski, M., 2020, "Materials, Well Designs, and Numerical Simulations for the Construction of a Low Loss Zone for Radiofrequency Downhole Heating. Paper presented at the *SPE Canada Heavy Oil Conference*, Virtual, September. https://doi.org/10.2118/199946-MS

Ovalles, C., Vaca, P., Okoniewski, M., Dieckmann, G., Pasalic, D., Dunlavey, J., 2018, "Numerical Simulation of Dielectric Heating in a Heavy Oil Reservoir Using a Shaped Dipole Antenna." Paper presented at the *SPE Canada Heavy Oil Technical Conference*, Calgary, Alberta, Canada, March https://doi.org/10.2118/189764-MS

PCL Project Detail, 2011, "Solar to Steam Demonstration Plant", Solar to Steam Demonstration Plant. Retrieved August 5, 2025.

Pérez, R., Osma, L., García-Duarte, H.A., 2024 "Combining Steam and Flue Gas as a Strategy to Support Energy Efficiency: A Comprehensive Review of the Associated Mechanisms," *ACS Omega*, 9, 15732–15743, https://doi.org/10.1021/acsomega.3c09889

Pérez, R.A., Rodriguez, H.A., Rendón, G.J., Plata, B.G., Salinas, L.M., Barbosa, C., García, L.E., Rojas, F.A., Orrego, J.A., León, L.J., St. Bernard, J.G., Manrique, E.J., 2022. "Optimizing Production Performance, Energy Efficiency and Carbon Intensity with Preformed Foams in Cyclic Steam Stimulation in a Mature Heavy Oil Field: Pilot Results and Development Plans." Paper presented at the *SPE Improved Oil Recovery Conference*, Virtual, April. https://doi.org/10.2118/209399-MS

Petroleum Watch, 2020, California Energy Commission, November. https://www.energy.ca.gov/sites/default/files/2020-12/2020-11%20Petroleum%20Watch_ADA.pdf. Retrieved November 1, 2025

Power Technology, 2018, "Belridge Solar Thermal Power Plant, California", California, US. Retrieved August 5, 2025.

Ramcharan, T., Hosein, R., Jupiter, A. 2016 "The Viability of Oil Extraction from Trinidad Tar Sands by Radio Frequency Heating - A Simulation Approach." Paper presented at the *SPE Trinidad and Tobago Section Energy Resources Conference*, Port of Spain, Trinidad and Tobago, June. https://doi.org/10.2118/180830-MS

Rassenfoss, S., 2012, "Oil Sands Get Wired - Seeking More Oil, Fewer Emissions," *J Pet Technol*, 64, 34–45. https://doi.org/10.2118/0912-0034-JPT

Redford, D.A., McKay, A. S., 1980, "Hydrocarbon-Steam Processes for Recovery of Bitumen from Oil Sands." SPE No. 8823. Paper presented at *the SPE/DOE Enhanced Oil Recovery Symposium*, Tulsa, Oklahoma, April. https://doi.org/10.2118/8823-MS

Rodriguez, F., Belhaj, H., Morales, R., Velasquez, R., AlDhuhoori, M., Alhameli, F. 2024 "Electrical/Electromagnetic Enhanced Oil Recovery for Unconventional Reservoirs: A Review of Research and Field Applications in Latin America." Paper presented at the *SPE Conference at Oman Petroleum & Energy Show*, Muscat, Oman, April. https://doi.org/10.2118/218665-MS

Rodriguez, F., Delamaide, E., Rousseau, D., Bekri, S., 2022, "Which is the Most Attractive IOR Method to Produce the Venezuelan Highly Viscous Oil Resources in the Energy Transition Era? A Comprehensive Review of Research and Field Applications." Paper presented at the ADIPEC, Abu Dhabi, UAE, October. https://doi.org/10.2118/211344-MS

Rosa, L., Davis, K.F., Rulli, M. C., D'Odorico, P., 2017, "Environmental Consequences of Oil Production from Oil Sands," *Earth's Future*, 5, 158–170, https://doi.org/10.1002/2016EF000484

Sadeghi, A., Boustani, A., Hassanzadeh. H., 2022 "Optimization of the Operating Envelope of a Hot-Solvent Injection Process for Bitumen Recovery," *SPE J.*, 27, 2268–2282. https://doi.org/10.2118/209592-PA

Safaei, M., 2019 "Estimation of the Life Cycle Greenhouse Gas Emissions of Bitumen-Derived Petroleum Fuels using Enhanced Solvent Extraction Incorporating Electromagnetic Heating (ESEIEH) and Toe-to-Heel Air Injection (THAI) Extraction Technologies", M. Sc. Thesis, Department of Mechanical Engineering, U. of Alberta, https://ualberta.scholaris.ca/server/api/core/bitstreams/7977af0c-982c-42cf-bb43-68ba74183d7d/content. Retrieved July 28, 2025.

Safaei, M., Oni, A.O., Gemechu, E.D., Kumar, A., 2019, "Evaluation of Energy and GHG Emissions' Footprints of Bitumen Extraction Using Enhanced Solvent Extraction Incorporating Electromagnetic Heating Technology", *Energy*, 186, 115854–115866. https://doi.org/10.1016/j.energy.2019.115854

Sahni, A., Kumar, M., Knapp, R. B. 2000, "Electromagnetic Heating for Heavy Oil Reservoirs", SPE No. 62550, presented at *SPE/AAPG Western Regional Meeting*, Long Beach, California, June 19–23. https://doi.org/10.2118/62550-MS

Sanchez, J., 2022, "Field Pilot of Steam Additives Technology to Reduce GHG Emissions Intensity in SAGD Final Non-Confidential Outcomes Report", December 22. https://www.eralberta.ca/wp-content/uploads/2023/10/CPC-Steam-Additives-Project-Final-Non-Confidential-Outcomes-Report-20230314-Updated-1.pdf. Retrieved June 2, 2025

Sander, P.R., Clark, G.J., Lau, E.C., 1991, "Steam-Foam Diversion Process Developed to Overcome Steam Override in Athabasca." Paper presented at the *SPE Annual Technical Conference and Exhibition*, Dallas, Texas, October. https://doi.org/10.2118/22630-MS

Sandler, J., Fowler, G., Cheng, K., Kovscek, A. 2012, "Solar-Generated Steam for Oil Recovery: Reservoir Simulation, Economic Analysis, and Life Cycle Assessment," Paper presented at the *SPE Western Regional Meeting*, Bakersfield, California, USA, March. https://doi.org/10.2118/153806-MS

Sandler, J., Fowler, G., Cheng, K., Kovscek, A.R., 2014, "Solar-Generated Steam for Oil Recovery: Reservoir Simulation, Economic Analysis, and Life Cycle Assessment," *Energy Convers. Manag.*, 77, 721–732. https://doi.org/10.1016/j.enconman.2013.10.026

Sleep, S., Dadashi, Z., Chen, Y., Brandt, A.R., MacLean, H.L., Bergerson, J.A., 2021, "Improving Robustness of LCA Results Through Stakeholder Engagement: A Case Study of Emerging Oil Sands Technologies," *J. Cleaner Prod.*, 281, 125277–125290. https://doi.org/10.1016/j.jclepro.2020.125277

Smalley, C., 2000. ""Heavy oil and Viscous Oil" Modern Petroleum Technology," Alan, R. A. D., ed., John Wiley & Sons Ltd., West Sussex, England, Volume 1. *Modern Petroleum Technology*, Volume 1, Upstream, 6th Edition|Wiley

Soiket, Md.I.H., Oni, A.O., Kumar, A., 2019 "The Development of a Process Simulation Model for Energy Consumption and Greenhouse Gas Emissions of a Vapor Solvent-Based Oil Sands Extraction and Recovery Process", *Energy*, 173, 799–808. https://doi.org/10.1016/j.energy.2019.02.109

Sresty, G.C., Dev, H., Snow, R.H., Bridges, J.E. 1986 "Recovery of Bitumen and Tar Sand Deposits with Radio Frequency Process", SPE No. 10229, *SPE Res. Eng.*, 1, 85–94. https://doi.org/10.2118/10229-PA

Stark, S. D., 2013, "Cold Lake Commercialization of the Liquid Addition to Steam for Enhancing Recovery (LASER) Process", IPTC 16795, presented at the *International Petroleum Technology Conference*, Beijing, China, 26–28 March. https://doi.org/10.2523/IPTC-16795-MS

Suncor Firebag, 2021, "2021 AER Scheme Performance Report Commercial Scheme Approval No. 8870", Reporting Period: January 1, 2021 to December 31, 2021. Retrieved June 11, 2025

Suranto, A., Bae, W., Permadi, A., 2015, "An Investigation of Hybrid Steam-Solvent Injection for Increasing Economy and Reducing CO_2 Emission," *Petrol Sci Technol*, 33(3), 302–310. https://doi.org/10.1080/10916466.2014.953683

Surmont In Situ Oil Sands Project, 2024, "Directive 054 Submissions & Annual OSCA Report", June, Surmont In Situ Oil Sands Project. Accessed May 27, 2025

Taylor, S.E., 2018, "Interfacial Chemistry in Steam-Based Thermal Recovery of Oil Sands Bitumen with Emphasis on Steam-Assisted Gravity Drainage and the Role of Chemical Additives," *Colloids Interfaces*, 2(2), 16. https://doi.org/10.3390/colloids2020016

Temizel, C., Irani, M., Canbaz, C.H., Palabiyik, Y., Moreno, R., Balikcioglu, A., Diaz, J.M., Zhang, G., Wang, J., Alkouh, A., 2018, "Technical and Economical Aspects of Use of Solar Energy in Oil & Gas Industry in the Middle East," Paper presented at the *SPE International Heavy Oil Conference and Exhibition*, Kuwait City, Kuwait, December. https://doi.org/10.2118/193768-MS

Temizel, C., Yegin, C., Hosgor, F.B., Aydin, H., Kabir, S., 2022, "Exploring Green Energy Applications in the Oil and Gas Industry," Paper presented at the *SPE Canadian Energy Technology Conference*, Calgary, Alberta, Canada, March. https://doi.org/10.2118/208900-MS

Trautman, M., Ehresman, D., Edmunds, N., Taylor, G., Cimolai, M., 2014a, "Effective solvent extraction system incorporating electromagnetic heating" US Patent No. 8,776,877.

Trautman, M., MacFarlane, B., 2014b. "Experimental and Numerical Simulation Results from a Radio Frequency Heating Test in Native Oil Sands at the North Steepbank Mine". Presented at the World Heavy Oil Conference, New Orleans LA, USA, Paper No. WHOC14-301.

Valera, C.A., Escobar, M.A., Iturbe, Y.J., 1999, "Use of Surfactants in Cyclic Steam Injection in Bachaquero-01 Reservoir," Paper presented at the *Latin American and Caribbean Petroleum Engineering Conference*, Caracas, Venezuela, April. https://doi.org/10.2118/54020-MS

Vega, G., Barrios, H., 2011, "Steam Injection Experiences in Heavy and Extra-Heavy Oil Fields, Venezuela", paper SPE-150283-MS presented at the *SPE Heavy Oil held in Kuwait City*, Kuwait, 12–14 December. https://doi.org/10.2118/150283-MS

Wang, Y., Ren, S., Zhang, L., Hu, C., 2018 "Energy Efficiency and Greenhouse Gas Emissions of Current Steam Injection Process and Promising Steam Based Techniques for Heavy Oil Reservoirs," *J. Pet. Sci. Eng.*, 166, 842–849. https://doi.org/10.1016/j.petrol.2018.03.094

Wang, Z., Gao, D., Liu, K., Tan, T., Wang, Z., Li, W., 2018, "Study on Radio Frequency Heating Pattern of Heavy Oil Reservoir Based on Multi-Antenna Configuration." Paper presented at the *SPE International Heavy Oil Conference and Exhibition*, Kuwait City, Kuwait, December. https://doi.org/10.2118/193774-MS

Wheeler, T.J., Dreher, W.R., Parsche, F.E., Trautman, M.A., 2016, "Enhanced recovery and in situ upgrading using RF", US Patent 9,453,400.

Wise, S., Patterson, C. 2016, "Reducing Supply Cost of EESIEH" Presented at the *SPE Canada Heavy Oil Technical Conference*, Calgary, Alberta, 7–9 June. SPE 180729. https://doi.org/10.2118/180729-MS

Yee, C.-T., Bharatha, S., McCaffrey, W.J., 2018, "Steam-solvent-gas process with additional horizontal production wells to enhance heavy oil/bitumen recovery" US. Patent No. 10,145,226, December 4.

Yee, C.T., Stroich, A., 2004, "Flue Gas Injection into a Mature SAGD Steam Chamber at the Dover Project (Formerly UTF)," *J. Can. Pet. Techno.* 43 No Pagination Specified. https://doi.org/10.2118/04-01-06

Yee, L., Chui, M., Roberts, R., Smit, S., 2025, "Future of energy and sustainability technologies", McKinsey technology trends outlook 2025 | McKinsey, page 99. Retrieved August 11, 2025

Yousefi, H., Abbaspour, A., Seraj, H., 2019 "Worldwide Development of Wind Energy and CO_2 Emission Reduction," *Env. Energy Econ. Res.*, 3(1), 1–9. https://doi.org/10.22097/eeer.2019.164295.1064

Zeidani, M., 2024 "Numerical and Mathematical Modeling of In-situ Reflux (ISR) Recovery Process" Doctoral thesis, University of Calgary, Calgary, Canada. Retrieved from https://hdl.handle.net/1880/118108

Zhang, S., Xiuluan, L., Wang, H., Sun, X., Jin, R., Gou, Y., Zhang, Z. Shen, D., 2021 "Experimental Study on the Process of eMSAGP", *Pet. Petro. Chem. Eng. J.*, 5(1), 1–13. https://doi.org/10.23880/ppej-16000255

Zhong, L., Wang, C., Liu, Y., Zhang, W., Han, X., Sun, Y., 2022 "Field Application of a Modular Multiple Thermal Fluid Generator for Heavy Oil Recovery," *J. Petrol. Explor. Prod. Technol.*, 12, 227. https://doi.org/10.1007/s13202-021-01376-y

Ziabakhsh-Ganjia, Z., Nicka, H.M., Donselaara, M.E., Bruhna, D.F., 2018, "Synergy Potential for Oil and Geothermal Energy Exploitation" *App. Ener.*, 212, 1433–1447. https://doi.org/10.1016/j.apenergy.2017.12.113

4 Emerging Technologies

In this chapter, several emerging technologies for the Enhanced Oil Recovery (EOR) of heavy crude oils (HCOs) will be discussed. These environmentally friendly technological routes are coming over the horizon with the potential to increase the oil rates and recovery factors, while simultaneously reducing greenhouse emissions, technology footprint, and water consumption. Many of these pathways for HCO production have Technology Readiness Levels (TRLs) of 6 or lower (refer to **Table 1.2** for definitions), indicating that they have been field tested or have been evaluated in the laboratory and/or successfully modeled (TRL $\leq$ 4). If the techno-economic assessments show promising results, these HCO-producing processes could advance to higher TRLs, with some potentially commercialized in the coming years.

The following sections will present several advanced, environmentally friendly technologies for HCO production. These include Nanotechnology and Nanomaterials (Section 4.1), Ionic Liquids and Deep Eutectic Solvents (Section 4.2), Sustainable Biomaterials (Section 4.3), Ultrasound Applications (Section 4.4), Subsurface Hydrogen Production (Section 4.5), and Switchable Surfactants (Section 4.6).

4.1 NANOTECHNOLOGY AND NANOMATERIALS

Over the past 30 years, the application of nanotechnology to the oil and gas value chain has experienced rapid growth (Agista 2018, Druetta 2018, Koa 2019, Osma 2019, Suwaid 2023, Gomaa 2024, Prada 2024, Kandiel 2025, and Jia 2025). In exploration, nanoparticles (NPs), nanofibers, and nanostructured materials have been utilized as sensors and other wellbore and reservoir measurement devices. During drilling and completion, silica, alumina, carbon, and metal-bearing NPs have been employed in drilling fluid formulations, cement acceleration, and thermal stability and mechanical properties enhancers. In stimulation and production, metal and metal oxide-containing NPs have been evaluated as additives for formation fracturing, emulsion stabilizers, viscosity enhancers, and reduction of interfacial tension (IFT). In refinery operations, nano-supported and metal-containing nanomaterials have been utilized as catalysts or catalyst precursors for the hydroconversion and hydrocracking of heavy fractions. They also serve to enhance water treatment, stabilize asphaltenes, and function as water-in-oil demulsifiers (Agista 2018, Druetta 2018, Koa 2019, Osma 2019, Gomaa 2024, Prada 2024, and Jia 2025).

Broadly, nanomaterials are materials that have at least one dimension—length, width, or height—in the range of 1–100 nanometers (nm). They often display unique optical, electrical, mechanical, or chemical properties that differ significantly from those of their bulk counterparts (Pacific 2025). Nanomaterials also offer several advantages, including high stability, ease of modifying their shape and size, higher thermal conductivity, enhanced chemical reactivity, and being environmentally friendly (Jia 2025). Thus, it is not surprising that these nanostructured materials

DOI: 10.1201/9781003635475-4

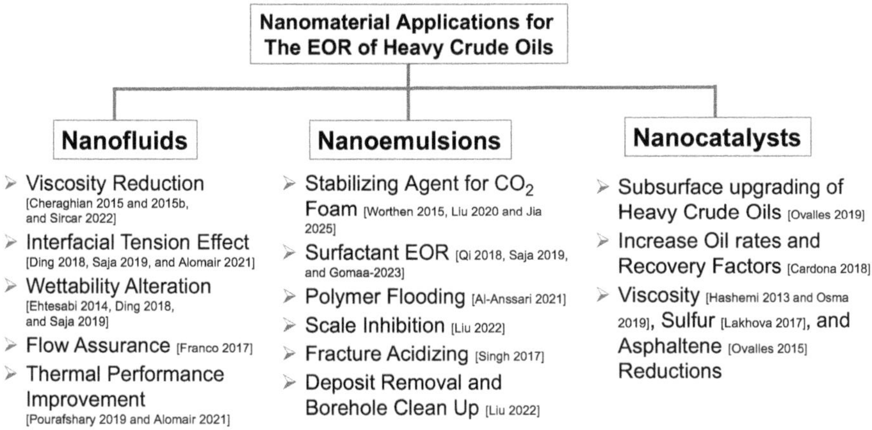

FIGURE 4.1 Examples of laboratory-scale nanomaterial applications for the Enhanced Oil Recovery (EOR) of heavy crude oils (HCOs) and a selected list of references.

have been employed for the EOR of HCO/B with increases in the recovery factor on 15–40% in comparison with conventional routes (Agista 2018, Druetta 2018, Koa 2019, Osma 2019, Gomaa 2024, Prada 2024, Kandiel 2025, and Jia 2025). **Figure 4.1** shows several examples of laboratory-scale applications for the EOR of HCOs and a selected list of references. In the first route, nanofluids are dispersed with NPs forming colloidal suspensions. They have been utilized for viscosity reduction (Cheraghian 2015a and 2015b, and Sircar 2022), interfacial tension (IFT) effect (Ding 2018, Saha 2019, and Alomair 2021), wettability alteration (Ehtesabi 2014, Ding 2018, and Saha 2019), flow assurance (Franco 2017), and thermal performance improvement (Pourafshary 2019 and Alomair 2021).

Conversely, nanomaterials could pose significant toxicity risks to humans and the environment due to their small size, high reactivity, and ability to penetrate biological systems (Sharifi 2012). Their effects vary widely depending on composition, shape, dose, and route of exposure. Therefore, their environmental and biological impacts must be evaluated on a case-by-case basis, and detailed ecotoxicological studies are necessary before drawing any conclusions.

Nanoemulsions are biphasic dispersions of two immiscible liquids: either water-in-oil (w/o) or oil-in-water (o/w) droplets stabilized by NPs, also known as Pickering emulsions (Agista 2018). As depicted in **Figure 4.1**, they have been studied in the laboratory scale for the EOR of HCOs as stabilizing agent for CO_2 foam (Worthen 2015, Liu 2020 and Jia 2025), surfactant-based EOR (Qi 2018, Saha 2019, and Gomaa 2024), polymer flooding (Al-Anssari 2021) as well as for scale inhibition (Liu 2022), fracture acidizing (Singh 2017), wax and asphalt deposit removal (Liu 2022), and borehole clean up (Liu 2022).

Nanocatalysts are metal-containing solids used for the chemical transformation and upgrading of HCO/B at downhole and refinery conditions (Ovalles 2019). As illustrated in **Figure 4.1**, they have been employed in the laboratory scale for subsurface upgrading of HCO (Ovalles 2019, Chapter 8, Section 8.2.4, and references therein), increase oil rates and recovery factors (Cardona 2018), viscosity (Hashemi 2013 and Osma 2019), sulfur (Lakhova 2017), and asphaltene reductions (Ovalles 2015).

4.1.1 Field Case Studies

Although there have been many experimental studies demonstrating the excellent potential of nanomaterials in EOR of HCO/B, its application in the field is limited to a few pilot tests (Kandiel 2025, and Jia 2025). **Table 4.1** presents several case studies reported in the literature (Zabala 2016, Wu 2023, Cárdenas 2024, Franco 2025). In entries 1 and 2, a nickel oxide-based nanofluid (supplied by Petroraza, https://petroraza.com/sitio/nanotecnologia/) was evaluated for two Colombian HCO reservoirs, Chichimene and Castilla (Zabala 2016). The authors conducted core flooding tests and found that the nanofluid was effective in removing asphaltenes, reducing viscosity, and restoring the core's original wettability. For the Chichimene field (**Table 4.1**, entry 1), two wells were stimulated by injecting 86 and 107 barrels (BBL) of the nanofluid. Instantaneous oil rate increases of 310 BBL (87% increase compared to pre-treatment) and 87 BBL were achieved (Zabala 2016).

Similarly, two Castilla oil wells (**Table 4.1**, entry 2) were stimulated with 200 and 150 BBL of the nanofluid. The authors found that the oil rates increased by 270 BBL (a 69% increase compared to pre-treatment) and 280 BBL, with BS&W reductions of ~11% (Zabala 2016). For Chichimene wells, the authors found a 98% reduction in viscosity within 9 days, and for the Castilla analogs, a 47% drop was measured in 30 days. The authors concluded that the nanofluid effect remained effective for 269 days, with 56 ppm retained in the reservoir at the end of this period, indicating long-term stability (Zabala 2016).

Wu and coworkers reported (see **Table 4.1**, entry 3) the use of a nanopolymer microsphere particle for the waterflooding block S301 of the Ansai Oilfield (Wu 2023). The authors injected 5 ton of 0.1–0.2% nanopolymer solution with a particle size of 100 nm. They found that the daily oil rate increased by 13% (from 346 to 394 BBLD) and that production increased by 15 MBBL over 12 months (Wu 2023). Cárdenas *et al.* (2024) (see **Table 4.1**, entry 4) investigated the use of a nanofluid in a Colombian reservoir. The stimulation process began with an initial phase of acid stimulation, followed by the flowback of the acid returns and the subsequent injection of nanofluids. The authors observed a significant increase in production, with rates rising from 10 to 467 BBL. Additionally, there was a 37% reduction in BS&W, dropping sharply from 99 to 62.5%. They also noted an increase in the API gravity of the HCO, which improved from 8.6° to 16°API over a period of three months (Cárdenas 2024).

Franco and coworkers (2025) (see **Table 4.1**, entry 5) utilized a nanofluid (Ni-Pd/Al_2O_3) to enhance the productivity of a Colombian heavy oil reservoir under cyclic steam stimulation (CSS). Initially, laboratory tests demonstrated the nanofluid's capability to reduce oil viscosity, improve oil mobility, and selectively interact with heavy oil fractions, such as resins and asphaltenes. Field trials conducted in two wells showed promising results, with average production increases of 97%. Well A produced an incremental yield of 11,966 BBL, while Well B produced 3,213 BBL. Furthermore, the API gravity increased from 11.6°API to 13.4°API, and viscosity decreased from

TABLE 4.1
Case Studies for the Use of Nanomaterials for the Enhanced Oil Recovery (EOR) of Heavy Crude Oils (HCOs)[a]

S. No.	Field (Years)	Operator (Country)	Heavy Oil Characteristics[b]	Nanofluid Composition[c]	Conditions[d]	Main Finding/Results	References
1	Chichimene (2014)	Ecopetrol (Colombia)	8°American Petroleum Institute (API) gravity, 1% water, and 16% C7-asphaltenes	Nickel oxide-based	86 and 107 BBL injected. 12-hour soaking	• Increase in oil rate up to 88% vs. pre-stimulation • 98% viscosity reduction within 9 days • No basic sediment and water (BS&W) change	Zabala (2016)
2	Castilla (2014)	Ecopetrol (Colombia)	12°API, 25% water, 8% C7-asphaltenes	Nickel oxide-based	200 and 150 BBL injected. 12–18-hour soaking	• Increase in oil rate up to 69% vs. pre-stimulation • 47% viscosity reduction in 30 days • 11% BS&W reduction	Zabala (2016)
3	Ansai (2020)	China National Petroleum Corporation (China)	18–22°API, NA	Nanopolymer microspheres, 100 nm	0.1–0.2% 5 ton injected	• Daily oil rate increased by 13% • Increase production of 15 thousand barrels (MBBL) in 12 months	Wu (2023)
4	Alpha (NA)	Ecopetrol (Colombia)	12–18°API, aveerage 9% C6-asphaltenes	NA, micron level	Acid injection, flowback, and nanofluid injection	• Oil rates increase from 10–467 BBL • Increase API gravity from 8.6° to 16°API in 3 months • Reduction in water cut by 37%	Cárdenas (2024)
5	N-A (2022–2023)	Ecopetrol (Colombia)	11.6°API 1.7% C6-asphaltenes	Ni-Pd/Al_2O_3 5.4 nm NiO 3.2 nm PdO	204 BBL injection followed by steam injection	• Production increased by 97% • Increase API gravity from 11.6° to 13.4°API • Viscosity decreased from ~40,000 to ~12,000 cP	Franco (2025)

[a] NA: Not available or not reported.
[b] HCO main characteristics.
[c] Main nanofluid compositions and particle size.
[d] Conditions of the field test.

39,987 cP (39.99 Pa·s) to 11,734 cP (11.73 Pa·s). These improvements were attributed to the interaction between the NPs and the oil, which enhanced crude oil mobility by altering the wettability of the reservoir rock. By modifying wettability, the oil flowed more easily through the reservoir, a crucial step in improving oil recovery efficiency. Additionally, the field trials indicated a 48% reduction in operational costs, mainly due to decreased steam generation and lower greenhouse gas (GHG) emissions compared to CSS. This highlights the environmental and economic benefits of using nanofluid technology in petroleum exploitation (Franco 2025).

Since 2005, the University of Calgary in Alberta, Canada, has developed a nanocatalytic-based process for the subsurface upgrading of HCO/B, known as ISUT®. This process aims to increase recovery, improve energy efficiency, and, at the same time, reduce environmental impact (Scott 2019, Moein-Elahi 2020, and Scheele-Ferreira 2023). ISUT® involves introducing a dispersed NiMo nanocatalyst directly into the reservoir. The nanocatalyst self-anchors to rock and sand surfaces, forming a permanent catalytic zone along the primary flow paths within the reservoir. Several applications have been proposed through a range of well configurations, ranging from vertical to horizontal and multilateral designs (Scott 2019 and Suarez-Suarez 2023). The catalytically upgraded oil exhibits improved mobility, enabling enhanced sweep and higher hydrocarbon recovery that might otherwise remain in place (Moein Elahi 2020).

Over the past four years, NanosTech in Alberta, Canada (https://www.nanos.tech/), has made significant progress toward deploying its technology in the field, focusing on numerical simulation and laboratory validation of ISUT® (Scheele-Ferreira 2023 and Suarez-Suarez 2023). Emission Reduction Alberta has approved a field test with a budget of CAD$ 5 million to demonstrate this technology at Christina Lake, Alberta (Emissions 2025). The objectives of this test are to increase production rates by up to 50% and achieve recovery of over 90% of the Original Oil in Place (OOIP). When used in conjunction with Steam-Assisted Gravity Drainage (SAGD), ISUT® can reduce water and steam requirements by up to 43% and lower GHG emissions by as much as 30% (Emissions 2025). As of the time of writing this monograph, the engineering phase of the field test is scheduled to begin in 2026.

4.1.2 Risks and Challenges

Using nanotechnology and nanomaterials for the EOR of HCO/B presents several risks and challenges to be considered and mitigated. The first issue is the synthesis, transportation, handling, and downhole injection of these materials. Nanofluids, nanoemulsions, and nanocatalysts are all fluids, so that the same procedures used for preparing and handling oilfield chemicals (e.g., demulsifiers and antifoaming) and solid suspensions (e.g., drilling muds) in surface facilities at or near the wellhead can be employed with success (Ovalles 2019). Another critical issue to consider is the cost of nanomaterials. During EOR processes, these materials are injected downhole and are not expected to be recovered. Thus, the need to use less expensive elements is paramount in reducing the operating cost of the process (Ovalles 2019, Hosny 2023, and Jia 2025).

NPs may be vulnerable to high temperatures, pressures, or salinity concentrations under downhole conditions (Druetta 2018, Prada 2024, and Jia 2025). Additionally, NPs dispersed in the porous media can be deposited or adsorbed onto the rock surface due to complex physicochemical interactions. This retention of NPs could result in pore blockage, potentially damaging the porosity and permeability of the rock (Prada 2024). The need for advanced numerical simulation tools is necessary to predict NP behavior and to optimize the EOR process.

In the following decades, environmental regulations are expected to significantly increase in Europe, the USA, and around the world. The use of nanomaterials in downhole applications could be classified as the release of hazardous chemicals into the environment unless comprehensive chemical evaluations and long-term studies are conducted to demonstrate otherwise (Kandiel 2025). In general, the toxicity of NPs depends on different parameters, including composition, source, dose, dimension, durability, mass, number, surface area, size, surface chemistry, aspect ratio, surface-coating, and function (Ahmadzadeh 2023). Thus, environmental concerns should be addressed early in the development of nanotechnology-based EOR processes, and appropriate measures must be implemented to prevent pollution and protect the biosphere.

The analysis in this section indicates that using nanomaterials for EOR for HCO/B production has shown a strong potential for success. This technology has been field tested in non-thermal (waterflooding) and thermal (CSS) processes under favorable or near-realistic conditions. Thus, a TRL level 5 can be assessed (see **Table 1.2** for definitions). Moreover, due to the significant increases in oil production rates and total accumulated oil output, its potential to reduce GHG emissions can be considered high (see **Figures 1.4** and **1.5**), provided that environmental and toxicological concerns are adequately addressed.

4.2 IONIC LIQUIDS AND DEEP EUTECTIC SOLVENTS

Over the past two decades, one of the emerging EOR technologies has been the use of ionic liquids (ILs) for chemical EOR. ILs are compounds formed by combining organic cations and inorganic/organic anions (Kharazi 2023 and Salam 2025). **Figure 4.2** illustrates the chemical structures of typical cations and anions utilized for the synthesis of ILs. The steric asymmetry resulting from the size and morphology of these ions leads to mismatches and the formation of unstable crystal lattices, which lowers the melting point. While these compounds are primarily salts, they differ from conventional molten salts in that they have a melting point ranging from −100°C to 200°C (Turosung 2017).

The main advantages of ILs include their thermal stability, resistance to salinity, higher relative interfacial activity compared to conventional amphiphiles, and relatively lower toxicity in comparison to commercial surfactants, alkalis, and polymers (Atilhan 2021, Kharazi 2023, Martinez-Palou 2023, and Salam 2025). In particular, ILs remain stable under SAGD or CSS high-temperature conditions, unlike some organic additives or polymeric materials. The application of ILs to the EOR of HCO/B has focused on several key areas: reducing IFT, altering rock wettability, preventing

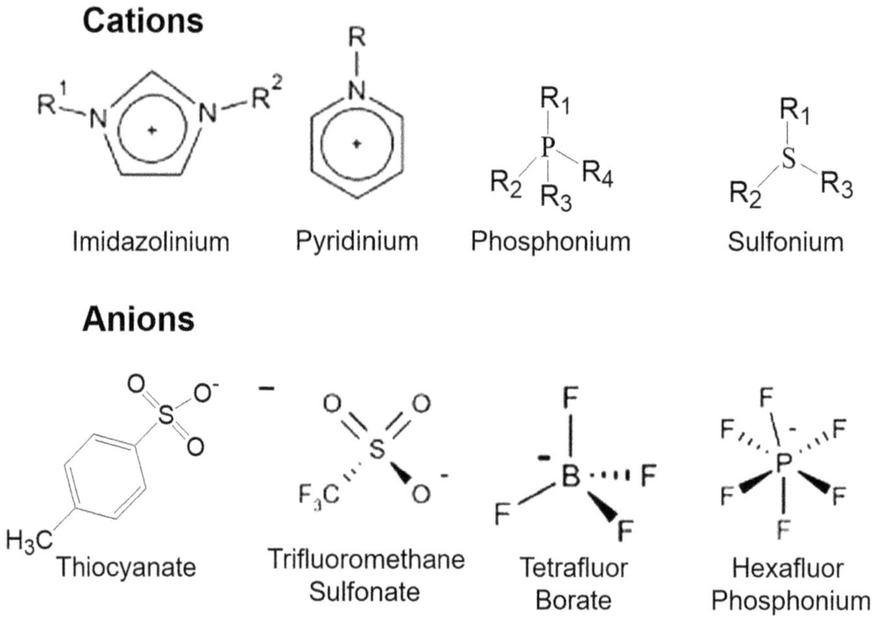

FIGURE 4.2 Chemical structures of typical ionic liquids (ILs).

asphaltene precipitation, and reducing viscosity. However, until now, all research related to EOR processes has been conducted at the laboratory level at reservoir temperatures, that is, non-thermal conditions (Turosung 2017, Atilhan 2021, and Martinez-Palou 2023).

A potential limitation of using ionic liquids in thermal EOR is that these substances typically stay in the liquid phase and do not travel with the steam. As a result, their penetration into the reservoir is limited and is not expected to exceed 30 m. Additionally, a significant drawback of using ILs in large quantities is their high cost. While they are relatively environmentally benign solvents, making them potential substitutes for volatile organic solvents, their high price has limited their application in the oil and gas industry (Turosung 2017, Alade 2023, and Martinez-Palou 2023). To overcome this challenge, several researchers have suggested the use of Deep Eutectic Solvents (DESs) (Koshekov 2020, Atilhan 2021, and Hamdi 2023). DES are mixtures of two or more components, typically a hydrogen bond donor (HBD) and a hydrogen bond acceptor (HBA). **Figure 4.3** illustrates the chemical structures of typical HBD and HBA utilized for the synthesis of DES. These components interact through hydrogen bonding to form a liquid at a temperature lower than the melting points of the individual substances. DESs have similar properties to ILs, such as viscosity, density, conductivity, IFT, and thermal stability. They are also inexpensive and easy to produce (Koshekov 2020 and Hamdi 2023). They are recognized as "green" solvents due to their low toxicity, biodegradability, and low volatility, making them a sustainable alternative to traditional solvents in various applications

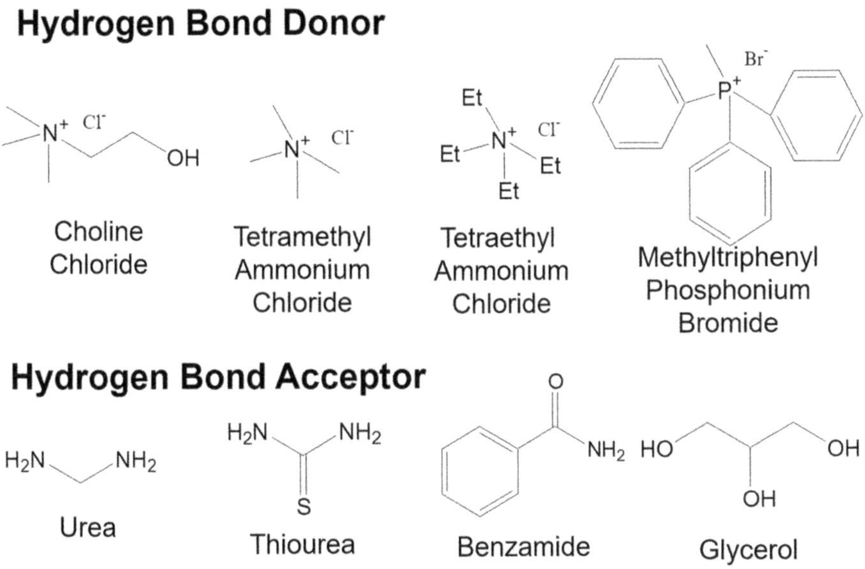

FIGURE 4.3 Chemical structures of typical Deep Eutectic Solvents (DESs).

such as extractions, electrochemistry, and food processing (Koshekov 2020 and Hamdi 2023). Some of the favorable properties include forming a stable emulsion, reducing interfacial tension, altering wettability, enhancing recovery, and preventing formation damage (Koshekov 2020 and Hamdi 2023).

At this moment, it is important to clarify that DES are not universally harmless or harmful; their safety and toxicity depend on the specific components and concentrations used (Hamdi 2020 and Alade 2023). While DES are often regarded as "green" alternatives that can effectively replace traditional harmful solvents, their environmental and biological impacts must be evaluated on a case-by-case basis. This is due to potential synergistic effects between components and a limited understanding of their overall ecotoxicological profiles (Wen, 2015). Therefore, detailed studies are necessary before drawing any conclusions (Abbott 2003, Wen 2015, Alade 2023, and Hamdi 2020).

Based on the discussion presented in this section, it can be concluded that the use of ILs for the non-thermal EOR of HCO/B is currently at a TRL 3 (proven at laboratory scale, see **Table 1.2**), as shown in **Figure 1.4**. There are some reports on the modeling and numerical simulations of IL flooding, which could suggest an advancement to TRL 4 (Hamdi, 2020 and Dahbag, 2016). Moreover, the potential reduction in GHG emissions can be attributed to the ability of ILs to enhance the efficiency of HCO recovery while decreasing the reliance on energy-intensive conventional production and upgrading processes. A medium potential for GHG emission reduction is registered in **Figure 1.4**. More details will be presented in Section 5.1.

In the next sections, several laboratory examples will be presented to illustrate the relatively high potential of ILs for the Reduction of Interfacial Tension

(Section 4.2.1), Wettability Change (Section 4.2.2), asphaltene inhibitors (Section 4.2.4), and Subsurface Upgrading: Permanent Viscosity Reduction (Section 4.2.4). The goal is not to provide a comprehensive literature review, but to highlight significant examples of the usefulness of IL for the EOR of HCO/B.

4.2.1 Reduction of Interfacial Tension

One of the principal advantages of ILs is their higher efficiency in reducing the IFT under high salinity conditions than their conventional counterparts with similar alkyl chain lengths and counter anions (Kharazi 2023 and Martinez-Palou 2023). This behavior is attributed to the bulkier head group, which gives IL-based surfactants greater hydrophobicity and easier transport to the interface, that is, more interfacial activity. This feature could make ILs a very useful alternative in surfactant flooding and in other Akali-Surfactant-Polymer (ASP) EOR processes (Hanamertani 2017, Atilhan 2021, Kharazi 2023, and Hamdi 2023). For instance, Hezave *et al.* found that the IFT between an Iranian crude oil and different solutions of the IL 1-dodecyl-3-methylimidazolium chloride decreased from 6.2 to 0.56 mN/m, as the NaCl concentration increased from 1 to 10 wt.% (Hezave 2013 and Hanamertani 2017). Conversely, for a conventional surfactant Ammoeng 102, the IFT was 4.8 mN/m at 10 wt.% NaCl. Thus, unlike the traditional amphiphiles, this IL-based surfactant was found to be more effective for higher saline formation water, which gives good potential for CEOR (Hezave 2013 and Hanamertani 2017).

Furthermore, the higher activity of the IL-based surfactants enables their use at low concentration, with the concomitant increase in the economic prospects of the IL EOR process. Laboratory research has shown that the IFT between Indian crude oil and a NaCl brine is lower when using a family of ILs based on lactam, ammonium, and hydroxylammonium, compared to non-ionic commercial surfactants (Sakthivel 2015 and Kharazi 2023). Additionally, ultra-low IFT values (<10–2 mN/m) can be achieved by using Zwitterionic, Catanionic, Multicationic, or Anionic IL-based surfactants. This wide variety of chemical structures highlights the versatility of ILs for the CEOR of HCO (Hanamertani 2017, Atilhan 2021, Kharazi 2023, and Hamdi 2023).

4.2.2 Wettability Change

Several authors have reported that IL-based surfactants are more efficient in wettability change from oil-wet to water-wet by altering rock/fluid interactions through electrostatic forces (Mohsenzadeh 2015, Nabipour 2017, Alade 2023, and Kharazi 2023). Contact angle measurements give a general qualitative indication of the wetting tendencies of the surfaces and fluids. In a typical example, Mohsenzadeh *et al.* found that the contact angle for an Omani HCO (16°API)/DES system changes from 28° (no DES) to 42° when using Choline Chloride/Glycerol (see **Figure 4.3**). This result indicates a shift from liquid-wetting to intermediately air-wetting conditions in the oil-air-rock system due to the presence of the DES (Mohsenzadeh 2015). This wettability alteration was attributed to changes in the surface energy, disruption of hydrogen bonding networks, and DES adsorption onto mineral surfaces (Kivela 2022 and Kumar 2025). In terms of oil recovery, using 50% DES/brine, the authors found ~44% recovery of the OOIP, whereas in the control experiment (no DES), 18% of the OOIP was obtained. This finding represents a 26% increase in recovery during DES flooding. The authors concluded that this DES showed suitable biodegradation and low toxicity (Wen 2015), with good potential for chemical EOR and minimal soil or water impact (Mohsenzadeh 2015).

Nabipour *et al.* (2017) compared the performance of an imidazolium IL-based surfactant (see **Figure 4.2**) with conventional products by using contact angle measurements. The authors found that IL-based amphiphiles are superior in changing the wettability from oil-wet to water-wet (Nabipour 2017). In general, the adsorption of IL-based surfactants on the reservoir surfaces converts oil-wet rocks to water-wet (Kharazi 2023). Cationic IL-based surfactants are particularly efficient in carbonate reservoirs, while anionic IL-based surfactants perform better in sandstone formations (Kharazi 2023). Furthermore, the structure and concentration of IL-based surfactants play a significant role in influencing rock wettability (Alade 2023). Surfactants with longer hydrocarbon chains and bulkier structures tend to show greater effectiveness (Atilhan 2021, Hamdi 2023, and Kharazi 2023).

Overall, applications of ILs for altering wettability are at a very preliminary stage (Alade 2023). Apart from understanding the mechanisms, the major impediment lies in the high cost and environmental impact due to toxicity, as mentioned earlier. However, IL-based surfactants have strong potential as alternatives to CEOR processes.

4.2.3 Asphaltene Inhibitors

ILs hold significant promise as asphaltene inhibitors and dispersants. They stabilize asphaltene aggregates, thereby preventing their precipitation and deposition in crude oil during production, transportation, and storage (Turosung 2017, Hamdi 2023, and Martinez-Palou 2023). The mechanism involves stabilizing asphaltene colloids through electrostatic interactions, π–π interactions, and hydrogen bonding. This process reduces the particle size of asphaltenes and lowers the viscosity of crude oil (Ghanem 2022 and Liu 2025). For instance, Ghanem and coworkers synthesized a series of imidazolium-based ILs and evaluated them as asphaltene dispersants using a viscometric technique (Ghanem 2022). They showed that the asphaltene onset precipitation was 28.5 vol.% in the absence of an additive (control experiment). When the IL-based additives are present, this value increased 42.8, 50, 78.5, and 64.3 vol.%, indicating an enhancement in the asphaltene stability (Ghanem 2022).

The same research group studied the effect of the imidazolium-based IL on the asphaltene stability by titration experiments (Ghanem 2023). They found that the asphaltene onset precipitation changed from 7.5 to 10.5 mL, 11, and 13.5 mL of added n-heptane when three different ILs were tested in the concentration range of 1000–2000 ppm. Additionally, the colloidal instability index of crude oil (see **eq. 4.1** (Xiong 2020), shifted from 0.92 (indicating unstable asphaltene) to 0.69 (indicating stable asphaltene) (Ghanem 2023).

$$\text{Collodial Instability Index} = \frac{\text{Saturates content} + \text{Asphaltenes content}}{\text{Aromatics content} + \text{Resins content}} \quad (4.1)$$

Liu and coworkers studied the inhibitory effects of three ILs and one commercial inhibitor on asphaltene deposition in crude oil (Liu 2025). The experimental results demonstrated that after inhibitor addition, the asphaltene precipitation onset point significantly increased from 34 to 52%, asphaltene aggregate particle size decreased from 1,082to 410nm, and turbidity values declined by 10–30%. This improvement was attributed to the ILs' ability to disrupt asphaltene molecular association via electrostatic interactions, steric hindrance, and hydrogen bonding, while simultaneously modifying the system's polarity, thereby effectively inhibiting asphaltene deposition (Liu 2025).

As a consequence of the IL-asphaltenes interactions, other authors have reported viscosity reductions upon blending the heavy crude oils with the ILs (Turosung 2017). For example, Saaid and coworkers found that a heavy oil (18°API) treated with 1 wt.% of a (1-butyl-3-methylimidazolium)(ClO_4) solution at 60°C for 72 hours led to a 27% reduction in the viscosity (Saaid 2014). Their findings also indicated that the presence of the IL significantly lowered the IFT of the oil-water interface from 21.2 to 18.3 mN/m, leading to an approximate 80% increase in the recovery factor (Saaid 2014).

Based on the laboratory examples published in the literature and discussed in this section, it can be concluded that ILs hold significant potential for inhibiting asphaltene deposition and viscosity reduction of HCO production, transportation, and storage. Further evaluations and field tests are necessary to move this technology to higher TRLs.

4.2.4 Subsurface Upgrading: Permanent Viscosity Reduction

Subsurface upgrading refers to the underground, *in situ*, or downhole recovery processes for HCO/B that result in permanent changes to the properties of these materials (Ovalles 2019). This process is characterized as a low-residue conversion or partial upgrading method. The primary goal of subsurface upgrading is to enhance the downhole mobility of HCO/B by reducing their viscosity. As expected, this improvement leads to increased oil rates and cumulative production, as well as better oil recovery and transport through pipelines. Additionally, subsurface upgrading enhances the value of the upgraded crude oils. The advantages and disadvantages of this process have been discussed in detail elsewhere (Ovalles 2019).

Table 4.2. shows selected examples of the use of IL for the upgrading of HCO/B at laboratory conditions. Fan *et al.* (entry 1) conducted the thermal upgrading of Liaohe crude oil in a batch reactor at very mild conditions (90°C) using the IL ($(Et)_3NH$) ($AlCl_4$) with and without transition metals, for example, Ni, Cu, or Fe for 4 days (Fan 2007). These experiments were carried out in the absence of a hydrogen source. The use of the IL led to a 44% reduction in viscosity, 49% removal of sulfur, and 18% conversion of asphaltenes (Fan 2007). Consistent with an acid-catalyzed process, after the upgrading reaction, the contents of saturates, aromatics, and resins increased, while the content of asphaltenes decreased. This change in composition led to a decrease in the average molecular weight and a reduction in the viscosity of the heavy oil (Fan 2007).

TABLE 4.2
Selected Literature Examples in the Use of Ionic Liquids (ILs) for the Upgrading of Heavy Crude Oil (HCO) at Laboratory Conditions

		Feed Characteristics			Reaction Conditions			Results			
S. No.	Catalyst	Crude Oil (Country)	°API	Viscosity (mPa·s @ °C)[a]	Temperature (°C)	Pressure (MPa)	Time (hour)	% Viscosity Reduction[b]	% DS[c]	% Asphaltene Conversion[d]	References
1	$(Et_3NH)AlCl_4$	Liaohe (China)	~15	420@50	90	ND	120	44	18	18	Fan (2007)
2	$(Et_3NH)AlCl_4Ni^{+2}$							65	85	43	
3	Fe-Mo-IL[e]	Offshore (Mexico)	13.5	1,444@37	350	10.8	72	97	50	68	Nares (2007)
4	(BMIM) $AlCl_4$[f]	Xinjiang (China)	21.7	7950@50	80	3.5	72	50	69	60	Fan (2009)
5	(BMIM) $FeCl_4$[g]	-(Egypt)	13.2	60@40	90	3.2	73	79	23	ND	Shaban (2014)
6	(HIMI)DBS[h]	-(Egypt)	15.9	1750@40	90	10	48	97	36	38	Alharthy (2022)

ND: Not disclosed or not reported.

[a] Viscosity (μ) in mPa·s measured at the given temperature.

[b] Percentage of decrease in viscosity of the upgraded crude oil vs. the feed calculated as $([\mu_{feed} - \mu_{product}] - \mu_{feed}) \times 100$.

[c] Percentage of desulfurization in the upgraded crude oil vs. the feed calculated as $((\%S_{feed} - \%S_{product}) - \%S_{feed}) \times 100$.

[d] Percentage of reduction of asphaltenes in the upgraded crude oil vs. the feed calculated as $([\%Asp_{feed} - \%Asp_{product}] - \%Asp_{feed}) \times 100$.

[e] The nature of the IL was not disclosed.

[f] BMIM = (1-butyl-3-methylimidazolium).

[g] 1-hexadecyl-1H-imidazol-3-ium 4-dodecylbenzenesulfonate.

Furthermore, Fan and coworkers reported improved results using a nickel-containing ionic liquid catalyst (see **Table 4.2**, entry 2). They reported a 65% viscosity reduction, 85% desulfurization, and 43% asphaltene conversion. The authors attributed the previous findings to a complex formed between the IL-Ni and the organic sulfur compounds present in the HCO. This complex weakens the C–S bonds with the concomitant sulfur removal (Fan 2007). However, no evidence was presented to support this mechanism.

Using more severe conditions (350°C, see **Table 4.2**, entry 3), Nares *et al.* obtained 97% viscosity reduction using a non-disclosed Fe and Mo-containing IL catalyst in 72-hour experiments (Nares 2007). The authors expected to achieve this temperature range by combining the catalytic heavy oil upgrading process with *In situ* Combustion. Unfortunately, no details were given on the integrated concept (Nares 2007).

In 2009, Fan and coworkers carried out the synthesis of the IL (1-butyl-3-methylimidazolium)($AlCl_4$) by the reaction of 1-methyl-imidazole with chloro-n-butane and aluminum chloride at 90°C for 20 hours (Fan 2009). The authors conducted thermal upgrading of Xinjiang crude oil (21.7°API) using (BMIM)($AlCl_4$) at 90°C, and 50% viscosity reduction, 66% desulfurization, and 59% asphaltene conversion were reported (see **Table 4.2**, entry 4) (Fan 2009).

Shaban and coworkers investigated the effects of reaction temperature and water amount on the upgrading of HCOs using (1-butyl-3-methylimidazolium)($FeCl_4$) (Shaban 2014). As shown in **Table 4.2**, entry 5, the authors reported 79% viscosity reduction and 23% sulfur removal at 90°C. The presence of water in the heavy oil had a detrimental effect on the upgrading and reduction in viscosity of heavy oils, which was attributed to catalysis poisoning (Shaban 2014).

On the contrary, Alharthy et al. (2022) utilized a series of acidic ionic liquids in the presence of 30% water to upgrade HCO. As indicated in **Table 4.2**, entry 6, the most promising results were obtained using the IL 1-hexadecyl-1H-imidazol-3-ium 4-dodecylbenzenesulfonate at 175°C for 48 hours. The authors reported a 94% reduction in viscosity, a 36% decrease in sulfur, and a 38% decrease in asphaltene content. They concluded that incorporating acidic ILs into the catalytic process is highly advantageous for upgrading HCOs due to their high catalytic effects, thermal stability, and surface activity (Alharthy 2022).

Based on the literature and evidence presented in this section, it can be concluded that the thermal upgrading of HCO/B can be effectively catalyzed by IL-containing catalysts at temperatures in the range of 90°C–175°C. This technology is at the laboratory level (TRL = 3) and has a medium to high potential for reducing GHG emissions, as illustrated in **Figure 1.5**. The use of IL-based catalysts in the presence of water at low temperatures is an intriguing approach for the thermal EOR of HCO/B and deserves further scrutiny.

4.3 SUSTAINABLE BIOMATERIALS

As discussed in Section 2.3, the injection of non-indigenous chemicals into down-hole environments may introduce potentially hazardous materials into heavy oil reservoirs. There is a risk of losing containment during workovers and production operations. Contaminants could potentially migrate toward nearby aquifers and/or surface water bodies through corroded or broken well casings, leaking tubing, improperly plugged wells, fractures, and faults (Nwosi-Anele 2019, Islam 2020, and Massarweh 2024). Over the last couple of decades, there has been a growing focus on utilizing environmentally benign and/or sustainable biomaterials for Chemical-Enhanced Oil Recovery (CEOR) of HCO/B, while simultaneously achieving higher oil recoveries and economic benefits (see Section 2.3 for more details).

Furthermore, recent technological advancements have reduced both chemical dosage requirements and CEOR costs, making the future outlook more promising (Mahdavi 2018, Islam 2020, and Rodriguez 2022). This section will examine two classes of bioderived materials. Those are microbial-derived surfactants, gases, and polymers (Section 4.3.1) and plant-based surfactants (Section 4.3.2). Additionally, laboratory and case studies and the environmental impacts of these technologies will be discussed.

4.3.1 Microbial-Derived Surfactants and Polymers

In Microbial-Enhanced Oil Recovery (MEOR), microorganisms and their nutrients are utilized to metabolically produce biosurfactants, biogases, and biopolymers to increase oil rates and cumulative crude oil production. For biosurfactants, several production mechanisms have been explored such as alteration of rock/sand wettability, o/w emulsification, and changes in the oil-water IFT (Islam 2020, and Massarweh 2024). Biogases, such as CH_4 and CO_2, can increase reservoir pressure and reduce oil viscosity, with the concomitant increasing oil production. Moreover, biopolymers can potentially improve oil recovery by altering the reservoir flow dynamics (Zhang 2020 and Pandey 2022).

Two EOR process modes have been reported using microbial-derived biomaterials. In the first one, *in situ* MEOR, the microorganism and nutrients (e.g., molasses, nitrates, and phosphates) are injected downhole to produce the biosurfactants, biogases, and biopolymers. In *ex situ* MEOR the microbial products are grown in bioreactors in the surface and then injected underground. Both processes have their advantages and disadvantages as well as their carbon footprint. A detailed comparison between those two MEOR processes is beyond the scope of this monograph, but it is available in the open literature (Zhang 2020, Islam 2020, Pandey 2022, and Massarweh 2024).

Broadly, microorganism-derived materials should not only effectively reduce the IFT between the aqueous and oil phases and alter wettability, but also demonstrate sufficient resilience to reservoir conditions, including pH, salinity, and temperature (Islam 2020 and Massarweh 2024). Safdel and coworkers evaluated 47 field tests across 21 countries, detailing the MEOR technologies employed, the names of the reservoirs or formations, the microorganisms used, the nutrients provided, and the specific effects observed in each case. (Safdel 2017). Several field and pilot cases have focused on biosurfactant-producing microorganisms, exploring how these organisms or their metabolic by-products—such as biosurfactants and biogases—can enhance crude oil mobilization and extraction. (Le 2015 and Massarweh 2024).

In a case study, Sun and workers conducted a microbial huff and puff pilot test using Bacillus spp. combined with indigenous bacteria to enhance heavy oil recovery in a reservoir located in central China (Sun 2017). The payzone had a thickness of 7.4 m (24.3 ft) and the OOIP was ~600,000 BBL of a HCO with a viscosity of 750 mPa·s. The reservoir had a temperature of 50°C, a porosity of 14.32%, and a permeability of 22 mD. Elemental analysis of the produced water indicated a significant shortage of nitrogen and phosphorus in the reservoir. As a result, nitrogen and phosphorus sources were added to the field test nutrients (Sun 2017).

Figure 4.4 shows the results of the pilot test on oil rates conducted before and after the huff and puff microbial EOR process (Sun 2017). As shown, the oil rate (light trace) increased from ~15 BBL/D prior to the microbial treatment to a maximum of ~30 BBL/D after just seven days of soaking (known in MEOR as "growing time"). Also, in **Figure 4.4**, the cumulative oil production (darker trace) showed a 13% increase compared to the prediction case (dashed line), highlighting the potential of MEOR to enhance oil production (Sun 2017).

Additionally, the authors reported that the viscosity of the produced crude oil decreased from 750 to 634 mPa·s (15% reduction) and the wax and asphaltene contents decreased by 12 and 17%, respectively (Sun 2017). These results were attributed to the production of biosurfactants and biogases by the bacterial streams, which can effectively biodegrade HCO components. Finally, the community structures changed significantly before and after microbial huff and puff, indicating the complexity of using MEOR in the field (Sun 2017). Unfortunately, no economic calculation nor carbon intensity results were provided.

Over the last decade, several biopolymers have been assessed for their application in MEOR, predominantly from the group of polysaccharides (Massarweh 2024). The most promising candidates among these include xanthan gum, welan gum, guar gum, scleroglucan, cellulose derivatives, chitin, and chitosan. Notably, xanthan and welan

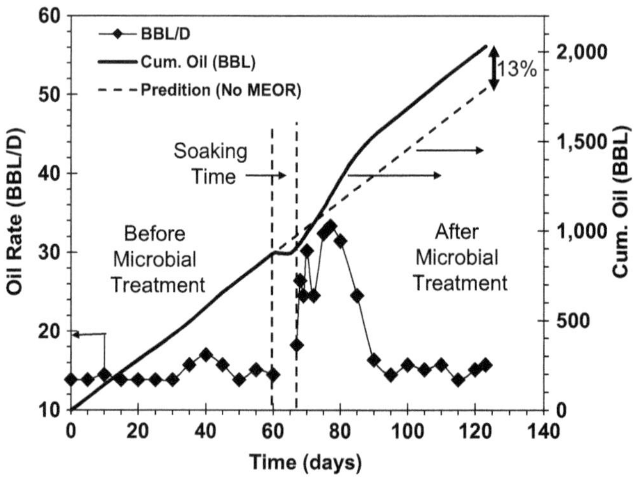

FIGURE 4.4 Oil rate (light trace), cumulative oil production (darker trace), and prediction case (no MEOR) before and after huff-n-puff microbial EOR process.

Data taken from (Sun 2017).

gums are produced through microbial fermentation processes involving the bacteria Xanthomonas campestris and Sphingomonas or Alcaligenes, respectively.

In a laboratory case study, Jang and coworkers evaluated the applicability of xanthan gum for the recovery of HCO (450 mPa·s at 25°C) (Jang, 2015). They compared its performance with that of hydrolyzed polyacrylamide (HPAM) at varying salinities, concentrations, and temperatures. In a laboratory flood test conducted with a 3 wt.% xanthan gum solution, a 7.2% increase in oil recovery was observed compared to the use of the HPAM solution. These experimental results indicated the potential feasibility of using xanthan gum flooding in high-salinity reservoir conditions (Jang, 2015).

As discussed in this section, microbial-derived surfactants and polymers offer a promising approach for sustainable oil production. However, there are significant challenges that must be adequately addressed. One of the main challenges of using biochemicals is their susceptibility to harsh reservoir conditions, such as high temperatures and salinity levels, which can impact their effectiveness and stability (Zhang 2020, Islam 2020, Pandey 2022, and Massarweh 2024). Moreover, despite over 50 field tests in MEOR and more than 500 related patents and publications, there remains uncertainty regarding the economics of these projects. Factors such as CAPEX, OPEX, the cost of bio-additives and nutrients, carbon intensity, environmental footprint, and water usage remain unclear (Quraishi 2021). Consequently, a TRL level 5 is currently assessed (see **Figure 1.4**).

MEOR minimizes the significant steam generation associated with thermal EOR, resulting in a less energy-intensive process overall. Moreover, certain microbes can convert heavier oil fractions into lighter ones. This conversion not only enhances the value of the crude oil through subsurface upgrading but also reduces the energy needed for downstream refining (Ovalles, 2019). Therefore, it is reasonable to assess a medium potential for reducing GHG emissions, as illustrated in **Figure 1.4**.

Finally, it is important to mention that there are alternative approaches reported in the literature regarding the use of microbial-derived surfactants and polymers for EOR. These approaches include Genetically Engineered Microbial-Enhanced Oil Recovery (GEMEOR) and Enzyme-Enhanced Oil Recovery (EEOR). The GEMEOR method involves using genetic engineering techniques to extract desirable features from multiple organisms to create more efficient strains, which results in improved oil recovery. In contrast, the EEOR technology employs enzymes, sometimes in conjunction with other enzymes or surfactants, to modify the IFT, wettability, mobility, and capillary action, ultimately leading to increased oil production (Patel 2015, Quraishi 2021, and Ke 2023). Currently, both methods are at the laboratory testing stages, with TRLs between 3 and 4.

4.3.2 Plant-Based Surfactants

In Section 2.4.2, biologically sourced solvents were utilized to reduce the viscosity of heavy oil mixtures and increase oil recovery. Additionally, in the pursuit of more affordable, higher-performing, and environmentally friendly materials, plant-based surfactants have emerged as a viable alternative to traditional chemical surfactants. The biomaterials have been effectively utilized in various sectors of the Oil & Gas industry, including wellbore and formation cleaning (Berry 2007), drilling fluids (Narehei 2012), viscosity reduction (Afra 2016), bitumen-based asphalts (Abe 2022), and EOR (Kamal 2017, Bachari 2019, Hama 2023, Andrade 2024, and Shaikhah 2024).

In Chemical EOR, the application of plant-based surfactants has been investigated using leaves and seeds of various plants, including Olive, Spistan, Prosopis, Mulberry, Ziziphus spina-Christi, Soapnut, and Chamomilla plants (Bachari 2019). Biosurfactants can be classified similarly to chemical surfactants, encompassing non-ionic, cationic, anionic, and amphoteric types (Bachari 2019). These biosurfactants offer numerous advantages over traditional options, such as biodegradability, low toxicity, availability of raw materials, resistance to harsh reservoir conditions, reduced water/oil IFT, altered wettability, and/or facilitated emulsification capabilities (Atta 2021, Andrade 2024, and Shaikhah 2024). Several studies have demonstrated that plant-based surfactants can improve ultimate oil recovery rates by 2 to 40% OOIP (Hama 2023).

However, plant-based surfactants suffer several challenges (Hama 2023). In some instances, the production of biomaterials can be more costly than traditional chemical surfactants (Shaikhah 2024). The costs associated can fluctuate based on various factors, including CAPEX and OPEX for extraction, purification, and formulation, production scale, and market variability (Hama 2023). To improve economic viability, researchers have utilized low-cost carbon and nitrogen sources, such as agro-industrial effluents, and have applied crude broth directly without the need for purification (Andrade 2024). Additionally, biosurfactants may exhibit compatibility issues with other chemicals used in EOR processes, potentially affecting their performance. Furthermore, the composition and properties of plant-based surfactants can differ significantly based on their source and extraction method. This variability can influence their effectiveness in reducing IFT and contact angle—both of which are essential for efficient EOR (Hama 2023).

The environmental impact of biosurfactants requires thorough evaluation (Shaikhah 2024). Although they are biodegradable and sourced from renewable resources, their production can still have ecological consequences. Long-term life cycle assessments of biosurfactants are necessary to verify their sustainability and environmental advantages. Additionally, since some biosurfactants cannot be used on their own in EOR without co-surfactants, it is essential to assess the compatibility of these combinations, along with their formulations, with existing policies, legislation, and regulatory requirements (Shaikhah 2024).

Although bio-derived surfactants have been demonstrated to be promising in laboratory and bench-scale trials (TRL = 3), there are still limited data on numerical simulation and field application, unlike traditional chemical surfactants (Kamal 2017, Bachari 2019, Hama 2023, and Shaikhah 2024). Pal *et al.* (2018) reported a laboratory case study that illustrates the potential of this technology. These authors synthesized, characterized, and evaluated an anionic biosurfactant from coconut oil for the EOR of Ahmedabad crude oil (API = 23.6, viscosity of 61.5 cSt at 30°C). The results indicated that biosurfactant solutions displayed good salt tolerance and long-term thermal stability. The IFT at the oil-aqueous interface was found to reduce to ultra-low values (0.004 mN/m) in the presence of 3.6 wt.% of NaCl.

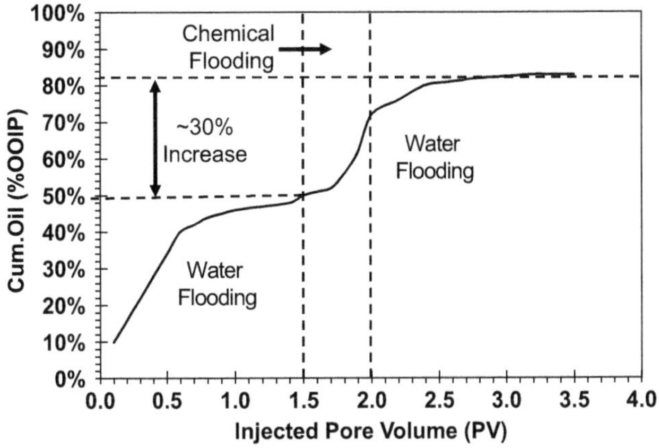

FIGURE 4.5 Cumulative oil recovery vs. the injected pore volume for the water and chemical sandpack flooding experiments using 0.8 wt.% of biosurfactant.

Data taken from (Pal *et al.* (2018)).

Additionally, the authors observed a change in the wettability of the rock from intermediate oil-wet to water-wet (Pal *et al.* (2018)).

Figure 4.5 shows cumulative oil recovery vs. injected pore volume for the water and chemical sandpack flooding experiments using 0.8 wt.% biosurfactant (Pal *et al.* (2018)). As shown, an additional recovery of almost 30% was obtained compared to the waterflooding case. The addition of 0.1–0.2 wt.% of partially hydrolyzed polyacrylamide slightly increased (~34%) the oil recovery (results not shown). The authors concluded that a coconut-derived anionic surfactant serves as a viable alternative for EOR (Pal *et al.* (2018)).

Based on the information present in this section, it can be concluded that the carbon intensity associated with the use of plant-based surfactants during CEOR operations is small to moderate. The main GHG emissions arise from the extraction, purification, formulation, and transportation of the biomaterials. Therefore, a conservative assessment indicates a medium potential for reducing GHG emissions, as illustrated in **Figure 1.4**. Additionally, as stated earlier, the TRL is rated at 3, meaning it has been proven effective at laboratory scales and through physical simulations. More details will be presented in Section 5.1.

4.4 ULTRASONIC APPLICATIONS

Sonic energy or ultrasound has been a very useful tool in enhancing the reaction rates for a variety of chemical systems (Suslick 1999, Thompson 1999, Grobas 2007, and Avvaru 2018) and has also been utilized for EOR (Sawarkar 2009, Shafiai 2020, Hamidi 2021, Agi 2022, Mierez 2024, and Gomaa 2024).

From the fundamental point of view, when a liquid is subject to sonication, the cavities within the liquid start oscillating in tune with the sound waves (Thompson 1999). This process, which involves the formation and dynamic behavior of these cavities, is known as cavitation. During the compression phase of the sound wave, the local pressure increases, while during the rarefaction phase, it decreases. If the local pressure drops significantly compared to the overall pressure, the cavity implodes. In cases where the pressure during the rarefaction cycle is low enough, localized areas of vaporized liquid can form small bubbles, referred to as cavitation bubbles. These bubbles are crucial to the phenomenon of acoustic cavitation and play a key role in sonochemical transformations. As the ultrasound wave cycles continue, the bubbles grow and contract, or implode during the subsequent compression phase. Each one of these imploding bubbles acts like a microreactor, with temperatures reaching over an estimated 5,000°C and pressures exceeding several hundred atmospheres. Therefore, the elevated local temperatures and pressures, along with extremely rapid cooling, create a unique environment for driving chemical reactions and inducing physical changes under extreme conditions. Although these phenomena occur at the micro-scale, the overall temperature and pressure of the reaction mixture and the vessel remain relatively mild (Suslick 1999, Thompson 1999, Sawarkar 2009, Grobas 2007, and Avvaru 2018).

Another energy efficient way of producing cavitation is through hydrodynamic cavitation (Avvaru 2018). This type of phenomenon occurs when a liquid flows through a constriction, such as a throttling valve, orifice plate, or venturi. A carefully designed system can create conditions for cavity collapse similar to those in acoustic cavitation, albeit with varying intensities and at significantly lower energy inputs than those required in sonochemical reactors. Several commercially available designs of hydrodynamic cavitation equipment utilize Bernoulli's principle. One practical application of this cavitation technology is the industrial-scale production of biodiesel (Avvaru 2018).

The applications of ultrasound energy and cavitation technologies to the petroleum industry are aimed at improving the well productivity by increasing the permeability of the near-wellbore region, subsurface upgrading (Ovalles 2019), and EOR (Avvaru 2018, Shafiai 2020, Hamdi 2021, Mierez 2024, and Gomaa 2024). Ultrasound has been shown to activate both chemical and physical processes in oil production, as demonstrated in laboratory and field tests (Agi 2022). This activation weakens the physical bonds at the boundary between rock pores and the fluids. Additionally, ultrasound alters the rheology of fluids by breaking the bonds between large molecules in viscous and heavy oil. As a result, more solid components, such as resins, paraffin, and asphaltene, become mobile, allowing them to separate from mineral deposits and facilitating deparaffinization (Agi 2022). Laboratory and field studies have shown that ultrasound treatment can nearly double oil production, depending on reservoir conditions, oil properties, and treatment duration. These topics will be covered in Sections 4.4.1 and 4.4.2, respectively.

Other examples have been reported in the literature, such as the breaking of the emulsion generated between invading oil-based mud filtrate and formation water, inorganic scale precipitation, plugging of pore throats by fines migration, and deposition of paraffin and asphaltenes (Avvaru 2018, Shafiai 2020, Hamdi 2021, Agi 2022, and Gomaa 2024); the authors stated that the main advantages of the ultrasonic technology are low energy consumption, the possibility to treat the near wellbore selectively, no physical harm to the well and casing, and it can be implemented in any reservoir type. Additionally, sonication routes are eco-friendly and economical EOR methods (Avvaru 2018, Shafiai 2020, Hamdi 2021, Agi 2022, and Gomaa 2024).

Conversely, a key disadvantage of downhole ultrasonic energy is that the vibrations do not travel long distances, limiting the effective range to less than 8–10 m. It has been observed that ultrasound intensity decreases inversely with the square of the distance from the transducer. This characteristic makes sonication an excellent technology for wellbore cleaning but less effective for deeper reservoir penetration, particularly beyond 10 m (Avvaru 2018, Shafiai 2020, Hamdi 2021, Agi 2022, and Gomaa 2024). Furthermore, higher oil viscosities can reduce the effectiveness of ultrasound treatment (Mullakaev 2015), which limits its application to HCO wells.

Other challenges include the fragility of the ceramic-based piezoelectric materials used in the ultrasound generators, the presence of heavy metals such as lead in the transducers, which can leach into the aquifers, and the need to correctly position the transducers to avoid loss of energy efficiency downhole. Additionally, while ultrasound technology has been used in onshore oilfields, its application in offshore oil production is limited due to the harsh conditions of underwater reservoirs (Avvaru 2018, Shafiai 2020, Hamdi 2021, Agi 2022, and Gomaa 2024).

This section, first, describes several typical examples of laboratory applications of ultrasound for EOR of HCOs. Second, a few case studies of field tests will be examined.

4.4.1 Laboratory Studies

Vibration-based oil recovery studies can be traced back to the 1950s, when oil recovery was observed to increase following earthquakes (Shafiai 2020). Since then, numerous studies have explored various aspects of the effects of ultrasonic waves in EOR. These studies have examined ultrasonic emitters or transducers, different types of fluids, and the effects of critical EOR and subsurface upgrading parameters, such as viscosity, permeability, pressure, temperature, and IFT (Naderi 2007, Ovalles 2019, and Li 2020). Several authors have reviewed the use of ultrasound energy or sonication for EOR at the laboratory level (Avvaru 2018, Shafiai 2020, Hamdi 2021, Agi 2022, and Gomaa 2024). This section will focus on its applications of HCOs by examining typical laboratory case studies.

Naderi and Babadagli from the University of Alberta investigated the effects of ultrasonic radiation on heavy oil recovery and multiphase flow through porous media, taking into account various rock wettabilities, oil viscosities, and initial water saturations (Naderi 2007). They utilized Berea sandstone cores and a heavy oil with a viscosity of 0.5 Pa·s (500 cP). Under imbibition conditions, they found that oil recovery reached 10% of the OOIP at an ultrasound intensity of 45 W/cm^2 (20 kHz); in contrast, it was only 0.2% without ultrasound. In contrast, a negligible effect of the ultrasound energy was found using a heavy oil with higher viscosity (1.6 Pa·s or 1,600 cP). The authors also noted that ultrasonic radiation had a more pronounced effect on oil recovery in oil-wet samples (Naderi 2007).

Paul and Davis from Mobil Corp. disclosed a downhole tool for HCO production that uses sonic energy and a metal-hydrogenation catalyst (Paul 1998). As shown in **Figure 4.6**, their invention comprises an acoustic transducer and an acoustic driver,

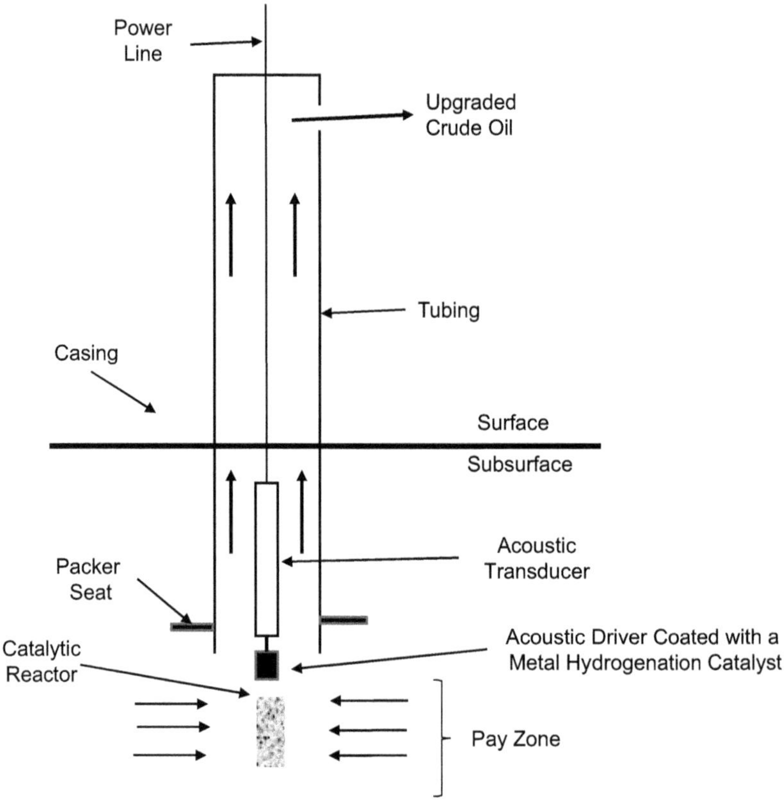

FIGURE 4.6 Schematic diagram of the downhole tool for the hydrotreating of heavy crude oil (HCO) in the presence of sonic energy and a metal hydrogenation catalyst.

to be positioned downhole just below the production tubing. The acoustic driver is coated with a Ni/Zn hydrogenation catalyst. The authors disclosed that a solid catalyst (e.g., Ni supported on ZnO) could also be utilized and placed in a small porous reactor bed located below the acoustic driver (see **Figure 4.6**). During production operation, water-containing HCO enters the casing and is subjected to sonic vibrations in the frequency range of 400–10 kHz (preferably about 1.25 kHz). At downhole temperatures and pressures, the authors proposed that water generates hydrogen via equation 4.2, which in turn is used to catalytically hydrotreat the HCO *in situ* (Paul 1998).

$$2H_2O \xrightarrow[\text{Catalyst}]{\text{Sonic Energy}} 2H_2 + O_2 \tag{4.2}$$

In laboratory experiments, the authors used a Battrum HCO emulsion containing about 40% v/v of water at the following conditions: 50°C, 100 psig argon pressure, 2.4 g nickel on zinc catalyst/140 ml HCO emulsion, sonic energy at a frequency of 1.25 kHz, weight hourly space velocity (WHSV) of 233 hour-1, and a reaction time of 15 m (Paul 1998). The results indicated that the asphaltene and resin contents decreased by 9% (from 16.19 to 14.69 wt.%) and 11% (from 41.38 to 36.71 wt.%), respectively. In the same experiment, the amounts of the aromatic and saturated fractions increased by 19% (from 30.95 to 36.88%) and 2% (from 11.48 to 11.71 wt.%), respectively. In a control experiment conducted without sonication, there were nearly no changes in the saturates, aromatics, resins, and asphaltenes (SARA) composition (Paul 1998).

Xu and Pu (2011) from China University of Petroleum (East China) conducted an experimental study on HCO upgrading using an undisclosed oil-soluble catalyst under ultrasonic energy. The results of displacement experiments showed that ultrasonic-assisted catalytic Aquathermolysis achieved 54% heavy oil recovery and 80.5% viscosity reduction. In contrast, the experiment using the catalyst alone achieved a recovery of 46% and a viscosity reduction of 72% (Xu 2011). The authors concluded that the combination of ultrasound and the catalyst had a synergistic effect, attributed to simultaneous viscosity reduction and upgrading of the heavy oil via Aquathermolysis. They also stated that sonication could provide energy for chemical reactions and improve catalyst activity, thus accelerating heavy oil cracking processes (Xu 2011).

4.4.2 Field Tests

The use of ultrasound in field tests has been tested in several parts of the globe since the early 2000s (Avvaru 2018, Shafiai 2020, and Agi 2022). Avvaru and coworkers reported that two ultrasonic downhole pieces of equipment had been developed and field tested (Avvaru 2018). These instruments have power ranging from 1 to 20 kW, working frequencies from 1 to 50 kHz, and withstand temperatures up to +150°C and pressures up to 50 MPa.

Figure 4.7 displays the oil production rates (in BBL/D) from selected field tests conducted before and after the application of ultrasound energy in oilfields located in China (Guo, 2004), Russia (Abramova, 2014; Mullakaev 2015), and the USA (Mullakaev 2015). The results indicate that ultrasound treatment can nearly double oil production, which aligns with the previously discussed laboratory experiments. Furthermore, a review of the literature reveals that water cut has decreased by 4% (from 32 to 28%), the success rate of the treatment ranges from 75 to 90%, and the beneficial effects can persist for a duration of 3–24 months (Agi, 2022).

Abramova and coworkers, as well as Mullakaev et al., have applied ultrasonic treatment to more than 100 wells in two regions of Russia, achieving increases in oil production rates between 40 and 100% (Abramova 2014; Mullakaev 2015). These researchers highlighted several advantages of ultrasonic EOR technology, including low energy consumption, selective reservoir treatment, no damage to the well or casing, and minimal environmental or human impact. In a subsequent article, Mullakaev and his colleagues (2017) noted that downhole sonication could also facilitate the introduction of various chemicals into the near-wellbore area and, further emphasizing its environmentally friendly nature (Mullakaev 2017).

Based on the laboratory and field data discussed in the section, it can be concluded that the application of ultrasonic energy to the EOR of HCOs is in TRL 6

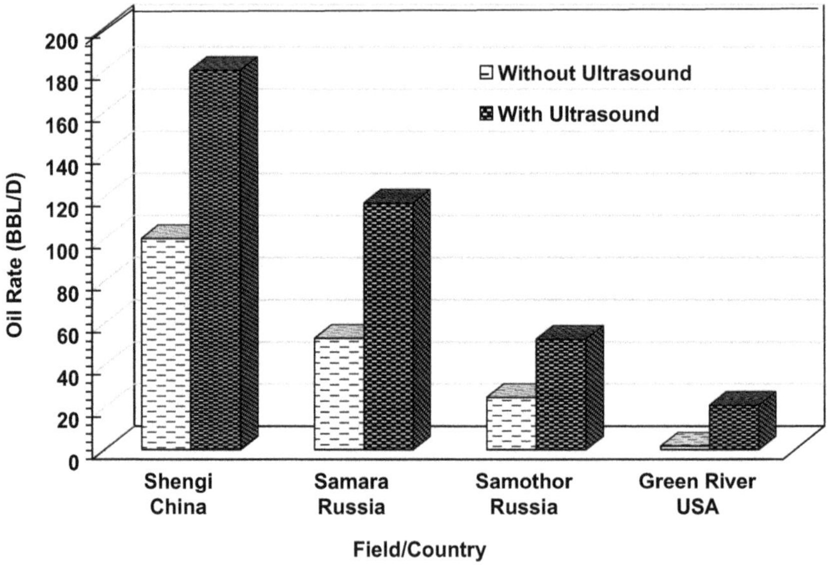

FIGURE 4.7 Oil rates (barrels per day [BBL/D]) for selected field tests before and after the application of ultrasound energy.

Data taken from (Guo 2004, Abramova 2014 and Mullakaev 2015).

(see **Table 1.2** for definitions). That is, it has been demonstrated under realistic conditions. The low energy consumption and small environmental impact reported by various authors suggest that this technology has a medium to high potential to reduce GHG emissions (see **Figure 1.5**). In comparison with other thermal processes and depending on the electricity cost, it can be speculated that the carbon intensity is around 20–60 kilograms of carbon dioxide equivalent removed for one US Barrel of petroleum (kg CO_2e/BBL).

4.5 SUBSURFACE HYDROGEN PRODUCTION

Over the past decade, emerging technologies have been developed to produce hydrogen directly from HCO/B reservoirs (Temizel 2022, Hu 2024, and Hanfi 2024). Although this route is not formally an EOR process, it offers an environmentally friendly approach to utilizing these abundant energy resources. These technological pathways have been tested in laboratory conditions and evaluated in the field. The approaches include *In situ* Combustion, gasification, and thermal- and electromagnetic-assisted catalytic heating. The basic principle behind these processes is to raise the reservoir temperature to initiate the chemical reactions necessary for hydrogen production. A selective membrane is then used to separate hydrogen from other gases, allowing H_2 to be collected at the surface. Meanwhile, by-products such as CO, CO_2, CH_4, and other hydrocarbon gases are recycled underground. This approach not only generates low-carbon intensity hydrogen but also minimizes environmental impact (Temizel 2022, Hu 2024, and Hanfi 2024).

4.5.1 Chemistry of Subsurface H_2-Production

The chemistry of subsurface hydrogen production depends on the method used to increase the reservoir temperature to start the H_2-generation reactions. Traditionally, *In situ* Combustion (ISC) projects involve injecting air or oxygen-enriched air underground (eq. 4.3), which results in some hydrogen production (Kapadia 2013, Ovalles 2019, Hanfi 2024, and Yuan 2025). The H_2-producing reaction includes steam reforming (eq. 4.4), the water-gas shift reaction (eq. 4.5), and gasification (eqs. 4.6 and 4.7). At the temperatures achieved during fire floods (>300°C), additional chemical processes can occur, such as thermal cracking with coke formation (eq. 4.8) and Aquathermolysis reactions (eq. 4.9). These various pathways happen simultaneously, making the *In situ* Combustion mechanism quite complex (Ovalles 2019, Chapter 9th).

$$\text{Heavy Oil Or Bitumens} + \text{Air or Oxygen} \xrightarrow{>300^\circ C} \text{Lighter Hydrocarbons} + CO + CO_2 + H_2O + \text{Heavy Fractions} + \text{Coke} \tag{4.3}$$

$$CH_4 + H_2O \rightarrow CO + 3H_2 \tag{4.4}$$

$$CO + H_2O \rightarrow CO_2 + H_2 \tag{4.5}$$

$$C(\text{coke}) + 2H_2O \rightarrow CO + H_2 \tag{4.6}$$

$$C + CO_2 \rightarrow 2CO \tag{4.7}$$

$$C_nH_m \rightarrow nC + (m/2)H_2 \tag{4.8}$$

$$RCH_2CH_2SHCH_3 + 2H_2O \rightarrow RCH_2 + CO_2 + H_2S + CH_4 + 2H_2 \tag{4.9}$$

In cases in which superheated steam is used to increase the reservoir temperature, the main pathways for H_2-generation are steam reforming (eq. 4.4) and gasification (eqs. 4.6 and 4.7). For instance, Surguchev *et al.* (2017) proposed a concept including the injection of steam and air along with a nano-sized catalyst for the subsurface hydrogen generation in hydrocarbon reservoirs. The role of the nano-sized catalyst was to lower the temperature needed to initiate the H_2-generation reactions. The authors validated this *in situ* hydrogen generation process through experimental and numerical modeling. The authors concluded by stating that their concept eliminates the energy use typically required for reducing CO_2 emissions and managing CO_2, such as in compression and transportation processes (Surguchev 2017).

Yan and coworkers reported the production of hydrogen from petroleum reservoirs using microwave heating (Yan 2023). The authors proposed that electromagnetic energy is transmitted into the reservoirs using a dipole antenna, and that heating efficiency and hydrocarbon reaction rate are improved by using Fe/Fe_3O_4 catalysts supported in fractures and rock pores away from the wellbore. The authors found that the shale rock achieved 100% methane conversion at 420°C. Demir et al. (2025)

applied a similar concept but in the absence of a catalyst to produce hydrogen from a 50:50 (v/v) mixture of crude oil and brine, oxygen, and using microwave energy. Results from five core experiments showed that hydrogen production occurred only when oxygen was present and was strongly influenced by heating rate and oil API gravity (ranging from 4.5 to 52). Microwave-assisted combustion consistently achieved earlier gas evolution and produced higher hydrogen yields, particularly from heavier oils such as bitumen (Demir 2025).

4.5.2 Field Test

Proton Technologies (https://www.proton-technologies.com/) conducted a field test, sponsored by the Saskatchewan Petroleum Innovation Incentive, using its patented technology for subsurface hydrogen generation (Gates 2022 and Proton 2024). The project took place at the Kerrobert Waseca Toe-to-Heel Air Injection (THAI) Project (Petrobank 2010). The target reservoir already had elevated temperatures of 100°C–150°C due to historical air injection through THAI and steam injection. The novel aspect of Proton's field demonstration was to optimize ISC and H_2-generation for the commercial production of hydrogen rather than hydrocarbon production. The schedule start date was January 2020 and concluded in June 2023, and covered the following deliverables (Proton 2024):

1. Hydrogen production field experiment on oil wells.
2. Drilling/completion/testing of an H_2 well.
3. Micro-scale surface membrane separation test.
4. Design and construction of H_2 compression, storage, and loading facilities.
5. Oxygen injection test.

For the hydrogen production test (step 1), air and steam were co-injected near the thickest part of the reservoir pay zone. Hydrogen was generated through the *In situ* Combustion (eq. 4.3), water-gas shift (eq. 4.5), and gasification (eqs. 4.6 and 4.7) reactions. The goal was to utilize the reservoir as the reaction vessel for hydrogen production and to validate the hypotheses and simulations that suggested it might be possible to produce hydrogen for under US$2.00/kg with a lower carbon intensity compared to blue hydrogen produced via Steam Methane Reforming (SMR). For comparison, the cost and carbon intensity of SMR typically range from US$1.50 to US$2.00 per kg and 9–12 kg CO_2/kg H_2, respectively (Proton 2024). Unfortunately, no data on the total H_2 per kg or the carbon intensity in kg of CO_2/BBL were reported.

Furthermore, Proton Technologies found that, over the course of the field test, there was no direct relationship between air injection rates and produced hydrogen concentrations. They were expected to be making 84 kg/d of hydrogen when the field test concluded. Although this goal was not achieved consistently due to the reservoir not reaching temperatures for optimal H_2 production, Proton did achieve a maximum of 120 kg/d and upward of 90 kg/d for four consecutive months (Proton 2024).

Proton Technologies concluded that future oxygen injection tests are critical for the next development phase of the project. These tests will be required to support the commercial viability of industrial-scale H_2-production, and the additional investment needed for commercial-scale oxygen production, large-scale gas separation, hydrogen purification, compression, storage, and transportation (Proton 2024).

The subsurface hydrogen generation via *In situ* Combustion and gasification, followed by membrane separation, poses significant challenges. One of the main hurdles in ISC is controlling, maintaining, and monitoring the burning front as it moves from the injector to the producer (Ovalles 2019). This task is not easy to accomplish, and it is the difference between success and failure. Even after 60 years of experimentation and field testing, *In situ* Combustion has a TRL levels between 7 and 8 (see **Table 1.4** for definitions), and successful deployment across multiple fields or wells worldwide (TRL 9) has not yet been fully realized.

Additionally, the combusted gases are collected by the producing well, where a Palladium-containing membrane selectively separates hydrogen from the other components and brings it to the surface. H_2-selective membranes are well-documented in the literature (Altaf 2024). However, there are significant uncertainties regarding their performance and reliability when treating combustion gases, particularly under downhole conditions. If the membrane fails, repairing the producing well could incur high costs, including lost production and rig expenses.

As mentioned, hydrogen flows to the surface in the producer, while the other gases are left behind. These gases contain not only CO and CO_2 but also methane, ethane, and minor amounts of C3-C10 hydrocarbons. Also, significant amounts of sulfur oxides (SO_x) and hydrogen sulfide (H_2S) are generated. Therefore, it is advisable to conduct geological and geomechanical tests to ensure that the gases trapped underground remain there and do not reach aquifers or the surface.

Based on the descriptions presented in this section, it can be concluded that the technologies for subsurface hydrogen production are at TRL 5, which indicates that the initial field tests were conducted under favorable conditions. Additionally, due to the capacity to trap GHG underground, the potential for emission reduction is assessed as medium to high, as illustrated in **Figure 1.5**.

4.6 SWITCHABLE SURFACTANTS

The literature contains numerous examples of switchable surfactants that can be activated by various stimuli, such as light, pH changes, or redox reagents (Liu 2006; Liang 2012). One notable type is long-chain acetamidines, which function as CO_2-triggered switchable surfactants. These materials exhibit surface activity in the presence of carbon dioxide but lose this activity when CO_2 is absent. These amphiphiles are also known as CO_2-responsive surfactants. **Figure 4.8** illustrates the use of N′-hexadecyl-N,N-dimethylacetamidine as a CO_2-switchable surfactant (Liang 2012). It has been reported that the application of CO_2-responsive surfactants facilitates carbon capture, utilization, and storage, contributing to the lowering of carbon intensity (Dong 2025).

In the past decade, CO_2-responsive surfactants have garnered considerable attention due to their unique ability to emulsify heavy oil underground while also demulsifying at the surface (Li 2024 and Dong 2025). EOR processes have specific requirements regarding surfactant surface activity as discussed in Section 2.3 "Chemical EOR" of this monograph. High surface activity is essential during oil

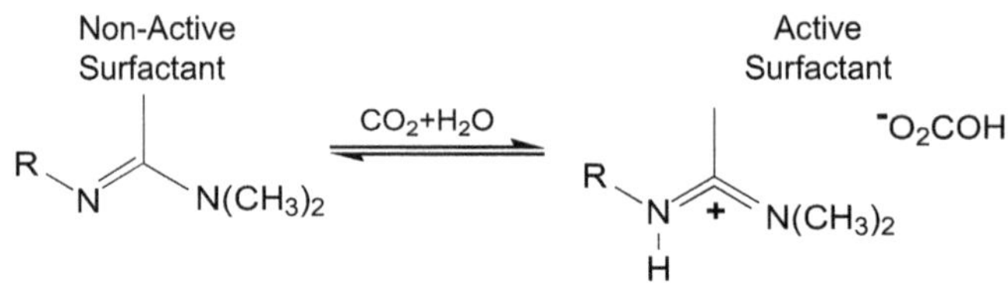

FIGURE 4.8 Typical example of N′-hexadecyl-N,N-dimethylacetamidine as a CO_2-switchable surfactant (Liang 2012). R: Long alkyl chain (C12 or C16, herein the latter presented).

production, whereas low or negligible surface activity is preferred during the heavy oil dehydration process in both upstream and downstream facilities. As previously mentioned, CO_2-switchable surfactants can adjust their surface activity by either adding or removing carbon dioxide from their aqueous solutions. This property makes them ideal for formulating switchable displacement systems, including CO_2-responsive foams, emulsions, hydrogel systems, oil sand washing, and oil-water separation (Li 2024 and Dong 2025). However, caution should be taken not to plug the reservoir by the produced water-in-oil emulsion (Zamani 2010).

Broadly, three types of CO_2-switchable surfactants have been reported in the literature (Li 2024 and Dong 2025). These are amidinoyl-responsive, guanidine- and amine-based surfactants. Among these, amine-based surfactants are considered the most promising for CO_2-activated applications. They can either be used in combination with common anionic amphiphiles available on the market or prepared independently by synthesizing amine surfactants through a series of steps (Dong 2025).

There are very few reports on the application of CO_2-switchable surfactants for the physical simulation of oil displacement in laboratories and field tests (Dong 2025). Li and coworkers reported that a "pseudo-Gemini" surfactant has been used in oil-well tests in Sinopec Shengli Oilfield since 2016 (Li 2024). Their results showed a significant enhancement in oil recovery in more than 90% of the tests (Li 2024). In 2024, PetroChina produced 10 ton of alkylamidine surfactant and plans to conduct field tests in Dagang and Liaohe Oilfields over the next two years (Li 2024).

In general, one of the main reasons for the limited commercialization of switchable surfactants is that their synthesis involves numerous complex steps, making large-scale industrial production difficult. Therefore, in the future, it will be necessary to optimize the synthesis procedure and explore more straightforward and high-yield synthetic methods to reduce costs (Liu 2024).

Based on the evidence presented in this section, it can be concluded that the chemical EOR technology using CO_2-switchable surfactants is at TRL 5 (successful initial field tests), with a medium potential to reduce GHG emissions, as shown in **Figure 1.4**.

4.7 CONCLUSION

- The use of nanomaterials for EOR for HCO/B has strong potential for success. This technology has been field tested in non-thermal (waterflooding) and thermal (CSS) processes under favorable or near-realistic conditions (TRL = 5).

- The utilization of ILs and DESs for the non-thermal EOR of HCO/B is currently at a TRL = 4. That is, it has been proven at the laboratory scale through physical and numerical simulations.
- Despite over 50 field tests in MEOR and more than 500 related patents and publications, there remains uncertainty about the economics of these projects, as well as their carbon intensity, environmental footprint, and water use. Consequently, a TRL level 5 is currently assessed.
- The application of plant-based surfactants during CEOR operations is at laboratory scale (TRL = 3) with a medium potential to reduce GHG.
- The ultrasonic energy technology of the EOR of HCO/B is at TRL level 6. The low energy consumption and small environmental impact reported by various authors suggest that this technology has a medium to high potential to reduce GHG emissions.
- The subsurface hydrogen production is at TRL level 5, which indicates that the initial field tests were conducted under favorable conditions. Additionally, due to the capacity to trap GHG underground, the potential for emission reduction is assessed as medium to high.
- The chemical EOR technology using CO_2-switchable surfactants is at TRL level 5 (successful initial field tests), with a medium potential to reduce GHG emissions.

REFERENCES

Abbott, A. P., Capper, G., Davies, D. L., 2003, "Novel Solvent Properties of Cholinechloride/Urea Mixtures," *Chem. Comm.*, 1, 70–71. https://doi.org/10.1039/B210714G

Abe, A. A., Rossi, C. O., Caputo, P., 2022, "Biomaterials and Their Potentialities as Additives in Bitumen Technology: A Review," *Molecules*, 27, 8826–8849. https://doi.org/10.3390/molecules27248826

Abramova, A., Abramov, V., Bayazitov, V., Gerasin, A., Pashin, D., 2014, "Ultrasonic Technology for Enhanced Oil Recovery," *Engineering*, 6, 177–184. https://doi.org/10.4236/eng.2014.64021

Afra, S., Nasr-El-Din, H., Socci, D., Cui, Z., 2016, "A Novel Viscosity Reduction Plant-Based Diluent for Heavy and Extra-Heavy Oil," Paper presented at the *SPE Improved Oil Recovery Conference*, Tulsa, Oklahoma, USA, April. https://doi.org/10.2118/179523-MS

Agi, A., Junin, R., Jaafar, M. Z., Sidek, M. A., Yakasai, F., Gbadamosi, A., Oseh, J., 2022, "Laboratory Evaluation to Field Application of Ultrasound: A State-of-the-Art Review on the Effect of Ultrasonication on Enhanced Oil Recovery Mechanisms," *J. Ind. Eng. Chem.*, 110, June, 100–119. https://doi.org/10.1016/j.jiec.2022.03.030

Agista, M. N., Guo, K., Yu, Z., 2018, "A State-of-the-Art Review of Nanoparticles Application in Petroleum with a Focus on Enhanced Oil Recovery", *Appl. Sci.*, 8(6), 871. https://doi.org/10.3390/app8060871, and reference therein.

Ahmadzadeh, S., Torkaman, P., Kalati, S. S., Jafari, A., Karimzadeh, R., 2023, "Environmental Effects of Enhanced Oil Recovery Methods", *J. Adv. Environ. Res. Tech.*, 1(1) 39–48.

Alade, O., Mohammed, I., Abdel-Azeim, S. Shakil-Hussain, S. M., Kamal, M. S., Mahmoud, M. Shehri, D. Al, Saibi, H., Adewunmi, A. A., Murtaza, M., Mahboob, A., 2023, "Review on Applications of Ionic Liquids (ILs) for Bitumen Recovery: Mechanisms, Challenges, and Perspectives", *Energy & Fuels*, 37(13), 8718–8738. https://doi.org/10.1021/acs.energyfuels.2c04175

Al-Anssari, S., Ali, M., Alajmi, M., Akhondzadeh, H., Manshad, A. K., Kalantariasl, A., Iglauer, S., Keshavarz, A., 2021, "Synergistic Effect of Nanoparticles and Polymers on the Rheological Properties of Injection Fluids: Implications for Enhanced Oil Recovery" *Energy Fuels*, 35, 7, 6125–6135. https://doi.org/10.1021/acs.energyfuels.1c00105

Alharthy, R. D., El-Nagar, R. A., Ghanem, A., 2022, "Laboratory Experiments on the In Situ Upgrading of Heavy Crude Oil using Catalytic Aquathermolysis by Acidic Ionic Liquid", *Materials*, 15, 5959–5977. https://doi.org/10.3390/ma15175959

Alomair, O., Alajmi, A., 2021, "A Novel Experimental Nanofluid-Assisted Steam Flooding (NASF) Approach for Enhanced Heavy Oil Recovery." *Fuel*, 313, 122691–122704. https://doi.org/10.1016/j.fuel.2021.122691

Altaf, C. T., Colak, T. O., Karagoz, E., Kurt, M., Sankir, N. D., Sankir, M., 2024, "A Review of the Recent Advances in Composite Membranes for Hydrogen Generation Technologies," *ACS Omega*, 9, 23138−23154. https://doi.org/10.1021/acsomega.4c00152

Andrade, A., Mehl, A., Mach, E., Couto, P., Mansur, C. R. E., 2024, "Application of Biosurfactants in Enhanced Oil Recovery Ex-Situ: A Review", *Brazilian J. Micro. Bio.*, 55, 3117–3139. https://doi.org/10.1007/s42770-024-01515-7

Atilhan, M., Aparicio, S., 2021, "Review on Chemical Enhanced Oil Recovery: Utilization of Ionic Liquids and Deep Eutectic Solvents," *J. Petrol. Sci. Eng.*, 205, 108746–108761. https://doi.org/10.1016/j.petrol.2021.108746

Atta, D. Y., Negash, B. M., Yekeen, N., Habte, A. D., 2021, "A State-Of-The-Art Review on the Application of Natural Surfactants in Enhanced Oil Recovery," *J. Mol. Liq.*, 321, 114888–114898. https://doi.org/10.1016/j.molliq.2020.114888

Avvaru, B., Venkateswaran, N., Uppara, P., Iyengar, S. B., Katti, S. S., 2018, "Current Knowledge and Potential Applications of Cavitation Technologies for the Petroleum Industry", *Ultrasonics Sonochemistry*, 42, April, 493–507. https://doi.org/10.1016/j.ultsonch.2017.12.010

Bachari, Z., Isari, A. A., Mahmoudi, H., Moradi, S., Mahvelati, E. H., 2019, "Application of Natural Surfactants for Enhanced Oil Recovery – Critical Review," *IOP Conf. Ser.: Earth Environ. Sci*, 221, 012039. https://doi.org/10.1088/1755-1315/221/1/012039

Berry, S. L., Boles, J. L., Cawiezel, K. E., 2007, "Evaluation of a Renewable, Environmentally Benign Green Solvent for Wellbore and Formation Cleaning Applications." Paper presented at the *International Symposium on Oilfield Chemistry*, Houston, Texas, U.S.A., February. https://doi.org/10.2118/106067-MS

Cárdenas, J. C., Jaimes, M., Dorado, R., Galvis, H., Conde, O. R., Prada, A., 2024 "Application of Nanofluids for Altering Wettability in Heavy Oil: A Field Test". Paper presented at the *SPE International Conference and Exhibition on Formation Damage Control*, Lafayette, Louisiana, USA, February. https://doi.org/10.2118/217924-MS

Cardona, L., Arias-Madrid, D., Cortés, F., Lopera, S., Franco, C. A., 2018, "Heavy Oil Upgrading and Enhanced Recovery in a Steam Injection Process Assisted by NiO- and PdO-Functionalized SiO2 Nanoparticulated" *Catalysts*, 8(4), 132–144. https://doi.org/10.3390/catal8040132

Cheraghian, G., 2015a, "An Experimental Study of Surfactant Polymer for Enhanced Heavy Oil Recovery Using a Glass Micromodel by Adding Nanoclay" *Pet. Sci. Technol.*, 33, 1410–1417. https://doi.org/10.1080/10916466.2015.1062780

Cheraghian, G., Khalilinezhad, S. S., 2015b, "Effect of Nanoclay on Heavy Oil Recovery During Polymer Flooding", *Pet. Sci. Technol.* 33, 999–1007. https://doi.org/10.1080/10916466.2015.1014962

Dahbag, M. S., Hossain, M. E., 2016, "Simulation of Ionic Liquid Flooding for Chemical Enhance Oil Recovery Using CMG STARS Software." Paper presented at the *SPE Kingdom of Saudi Arabia Annual Technical Symposium and Exhibition*, Dammam, Saudi Arabia, April 2016. https://doi.org/10.2118/182836-MS

Demir, A. B., Albabawat, Z. W., Hajiyev, M., Karabayanova, L., Ye, H., Hascakir, B., 2025, "Hydrogen at the Source: Unlocking Clean Energy from Bitumen with Advanced Thermal Recovery," Paper presented at the *SPE Energy Transition Symposium*, Sugar Land, Texas, USA, September. https://doi.org/10.2118/228353-MS

Ding, Y., Zheng, S., Meng, X., Yang, D., 2018, "Low Salinity Hot Water Injection with Addition of Nanoparticles for Enhancing Heavy Oil Recovery under Reservoir Conditions," Paper presented at the *SPE Western Regional Meeting*, Garden Grove, California, USA, April. https://doi.org/10.2118/190132-MS

Dong, B., Xu, Q., Liu, J., Du, S., Luo, W. Wu, W., Zou, X., 2025 "Liang, S. Research of CO2-Responsive Surfactants for Enhanced Oil Recovery: Review and Outlook", *Energies*, 18, 574–598. https://doi.org/10.3390/en18030574

Druetta, P., Raffa, P., Picchioni, F., 2018, "Plenty of Room at the Bottom: Nanotechnology as Solution to an Old Issue in Enhanced Oil Recovery" *Appl. Sci.*, 8, 2596. https://doi.org/10.3390/app8122596, and reference therein.

Ehtesabi, H., Ahadian, M. M., Vahid Taghikhani, V., Ghazanfari, M., 2014, "Enhanced Heavy Oil Recovery in Sandstone Cores Using TiO2 Nanofluids", *Energy & Fuels*, 28 (1), 423–430. https://doi.org/10.1021/ef401338c

Emissions Reduction Alberta, 2025, "Field Validation for In-Situ Upgrading Technology (ISUT™)". https://www.eralberta.ca/projects/details/field-validation-for-in-situ-upgrading-technology-isut/, Retrieved Noveber 7, 2025

Fan, H.-F., Li, Z.-B., Liang, T., 2007, "Experimental Study on using Ionic Liquids to upgrade heavy oil", *J. Fuel Chem. Tech.*, 35 (1), 32–35. https://doi.org/10.1016/S1872-5813(07)60009-7

Fan, Z.-X., Wang, T.-F., He, Y.-H., 2009, "Upgrading and Viscosity Reducing of Heavy Oils by (BMIM)($AlCl_4$) Ionic Liquid", *J. Fuel Chem. Tech.*, 37 (6), 690–693. https://doi.org/10.1016/S1872-5813(10)60015-1

Franco, C. A., Franco, C. A., Salinas, L. M., Alzate, L. G., Molina, D., Rendón, G. J., Obregón, C. C., Lopera, S. H., Medina, O. E., Cortés, F. B., 2025, "First field application of functionalized nanoparticles-based nanofluids in thermal enhanced oil recovery: From laboratory experiments to cyclic steam stimulation process" *Fuel*, 382, Part B, 133736–133751. https://doi.org/10.1016/j.fuel.2024.133736

Franco, C. A., Zabala, R., Cortés, F. B., 2017, "Nanotechnology Applied to the Enhancement of Oil and Gas Productivity and Recovery of Colombian Fields," *J. Pet Sci. Eng.*, 157, 39–55. https://doi.org/10.1016/j.petrol.2017.07.004

Gates, I. D., Wang, J., 2022, "In-situ process to produce hydrogen from underground hydrocarbon reservoirs", US Patent No. 11,530,603, Dec. 20. US Patent No. 11530603

Ghanem, A., Alharthy, R. D., Desouky, S. M., El-Nagar, R. A., 2022, "Synthesis and Characterization of Imidazolium-Based Ionic Liquids and Evaluating Their Performance as Asphaltene Dispersants." *Materials*, 15, 1600–1636. https://doi.org/10.3390/ma15041600

Ghanem, A., Nessim, M. I., Khalil, N. A., El-Nagar, R. A., 2023 "Imidazolium-Based Ionic Liquids as Dispersants to Improve the Stability of Asphaltene in Egyptian Heavy Crude Oil," *Sci. Rep.*, 13, 17158–17173. https://doi.org/10.1038/s41598-023-44237-w

Gomaa, S., Salem, K. G., El-Hoshoudy, A. N., 2024, "Enhanced Heavy and Extra Heavy Oil Recovery: Current Status and New Trends", *Petroleum*, 10 (3), 399–410. https://doi.org/10.1016/j.petlm.2023.10.001

Grobas, J., Bolivar, C., Scott, C. E., 2007, "Hydrodesulfurization of Benzothiophene and Hydrogenation of Cyclohexene, Biphenyl, and Quinoline, Assisted by Ultrasound, Using Formic Acid as Hydrogen Precursor", *Energy & Fuels*, 21, 19–22. https://doi.org/10.1021/ef0603939

Guo, X., Du, Z., Li, G., Shu, Z., 2004, "High Frequency Vibration Recovery Enhancement Technology in the Heavy Oil Fields of China." Paper presented at the *SPE International Thermal Operations and Heavy Oil Symposium and Western Regional Meeting*, Bakersfield, California, March 2004. https://doi.org/10.2118/86956-MS

Hama, S. M., Manshad, A. K., Ali, J. A., 2023, "Review of the Application of Natural Surfactants in Enhanced Oil Recovery: State-of-the-Art and Perspectives", *Energy Fuels*, 37, 10061–10086. https://doi.org/10.1021/acs.energyfuels.3c01347

Hamdi, Z., Koshekov, S., Bataee, M., 2023, "Potential of Deep Eutectic Solvents in the Upstream Oil and Gas Industry." Paper presented at the *SPE EuropEC - Europe Energy Conference featured at the 84th EAGE Annual Conference & Exhibition*, Vienna, Austria, June. https://doi.org/10.2118/214427-MS

Hamdi, Z., Kumaran, P. N., Talabi, O., 2020, "Investigation and Optimization of Deep Eutectic Solvent in Enhanced Oil Recovery Using Numerical Simulation." Paper presented at the *Abu Dhabi International Petroleum Exhibition & Conference*, Abu Dhabi, UAE, November. https://doi.org/10.2118/202633-MS

Hamidi, H., Haddad, A. S., Otumudia, E. W., Rafati, R., Mohammadian, E., Azdarpour, A., Giles Pilcher, W. G., Fuehrmann, P. W., Sosa, L. R., Cota, N., García, D. C., Ibrahim, R. M., Damiev, M., Tanujaya, E., 2021, "Recent Applications of Ultrasonic Waves in Improved Oil Recovery: A Review of Techniques and Results", *Ultrasonics*, 110, 106288. https://doi.org/10.1016/j.ultras.2020.106288

Hanamertani, A. S., Pilus, R. M., Irawan, S., 2017, "A Review on the Application of Ionic Liquids for Enhanced Oil Recovery." In: Awang, M., Negash, B., Md Akhir, N., Lubis, L., Md. Rafek, A. (eds) *ICIPEG 2016*. Springer, Singapore. https://doi.org/10.1007/978-981-10-3650-7_11

Hanfi, M. A., Alade, O. S., Tanimu, A., Mahmoud, M., Alarifi, S. A., 2024, "Catalytic and Noncatalytic in Situ Hydrogen Production from Heavy Oil: A Review of Experimental Studies" *ACS Omega*, 9, 50118–50133. https://doi.org/10.1021/acsomega.4c07383

Hashemi, R., Nassar, N. N., Pereira-Almao, P., 2013 "Enhanced Heavy Oil Recovery by in Situ Prepared Ultradispersed Multimetallic Nanoparticles: A Study of Hot Fluid Flooding for Athabasca Bitumen Recovery", *Energy Fuels*, 27(4), 2194–2201. https://doi.org/10.1021/ef3020537

Hezave, A. Z. Dorostkar, S. Ayatollahi, S. Nabipour, M., Hemmateenejad, B., 2013 "Investigating the Effect of Ionic Liquid (1-dodecyl-3-methylimidazolium chloride ((C12mim)(Cl))) on the Water/Oil Interfacial Tension As a Novel Surfactant," *Coll. Surf. A: Physicochem. Eng. Aspects*, 421, 63–71. https://doi.org/10.1016/j.colsurfa.2012.12.008

Hosny, R., Zahran, A., Abotaleb, A., Ramzi, M., Mubarak, M. F., Zayed, M. A., Shahawy, A. E., Hussein, M. F., 2023, "Nanotechnology Impact on Chemical-Enhanced Oil Recovery: A Review and Bibliometric Analysis of Recent Developments," *ACS Omega*, 8, 46325–46345. https://doi.org/10.1021/acsomega.3c06206

Hu, Q., Yan, L., Cheng, F., 2024, "*In-situ* Hydrogen Production from Petroleum Reservoirs and the Associated High Temperature Hydrogen Attack: A Review," *Int. J. Hydrogen Energy*, 95, 1038–1051. https://doi.org/10.1016/j.ijhydene.2024.11.280

Islam, M. R., 2020 "Economically and Environmentally Sustainable Enhanced Oil Recovery" Wiley, Hoboken, NJ 07030, USA and Scrivener, Beverly, MA 01915, USA, ISBN 978-1-119-47909-3

Jang, H. Y., Zhang, K., Chon, B. H., Choi, H. J., 2015 "Enhanced Oil Recovery Performance and Viscosity Characteristics of Polysaccharide Xanthan Gum Solution," *J. Ind. Eng. Chem.*, 21, 741–745. https://doi.org/10.1016/j.jiec.2014.04.005

Jia, H., Kang, Y., 2025, "A Comprehensive Review on Application and Perspectives of Nanomaterials in Enhanced Oil Recovery," *Energy & Fuels*, 39 (6), 2916–2942, https://doi.org/10.1021/acs.energyfuels.4c04848, and references therein.

Kamal, M. S., Hussein, I. A., Sultan, A. S., 2017 "Review on Surfactant Flooding: Phase Behavior, Retention, IFT, and Field Applications", *Energy & Fuels*, 31 (8), 7701–7720. https://doi.org/10.1021/acs.energyfuels.7b00353

Kandiel, Y. E., Attia, G. M., Metwalli, F. I., Khalaf, R. E., Mahmoud, O., 2025 "Nanoparticles in Enhanced Oil Recovery: State-Of-The-Art Review," *J. Pet. Expl. Prod. Tech.*, 15, 66–113. https://doi.org/10.1007/s13202-025-01965-1

Kapadia, P. R., Wang, J., Kallos, M. S., Gates, I. D., 2013, "Practical Process Design for *In Situ* Gasification of Bitumen" *Applied Energy*, 107, 281–296. https://doi.org/10.1016/j.apenergy.2013.02.035

Ke, C. Y., Sun, R., Wei, M. X., Yuan, X. N., Sun, W. J., Wang, S. C., Zhang, Q. Z., Zhang, X. L., 2023, "Microbial Enhanced Oil Recovery (MEOR): Recent Development and Future Perspectives", *Crit. Rev. in Biotechnol.*, 44 (6), 1183–1202. https://doi.org/10.1080/07388551.2023.2270578

Kharazi, M., Saien, J., 2023, "Chemical Enhanced Oil Recovery Using Ionic Liquid-Based Surfactants", Ed. Owoseni, O. IntechOpen, October 12, https://www.intechopen.com/chapters/88026#

Kivela, H., Salomaki, M., Vainikka, P., Mäkila, E., Poletti, F., Ruggeri, S., Terzi, F., Lukkari, J., 2022, "Effect of Water on a Hydrophobic Deep Eutectic Solvent," *J. Phys. Chem. B*, 126, 513−527. https://doi.org/10.1021/acs.jpcb.1c08170

Koa, S., Huhb, C., 2019, "Use of Nanoparticles for Oil Production Applications", *J. Pet. Sci. Eng.*, 172, 97–114, https://doi.org/10.1016/j.petrol.2018.09.051, and reference therein.

Koshekov, S., 2020 "Evaluating Deep Eutectic Solvent as a Novel Enhanced Oil Recovery Method", M.S. Thesis, Politecnico di Torino-Heriot Watt University, July. Deep Eutectic Solvent for EOR. Retrieved November 3, 2025

Kumar, A., Medha, S., Chakraborty, D., Kundu, D., Khan, S., 2025, "Enhanced Oil Recovery Promoted by Aqueous Deep Eutectic Solvents on Silica and Calcite Surfaces: A Molecular Dynamics Study", *Phys. Chem. Chem. Phys.*, 27, 9573–9589. https://doi.org/10.1039/D4CP04888A

Lakhova, A., Petrov, S., Ibragimova, D., Kayukova, G., Safiulina, A., Shinkarev, A., Okekwe, R., 2017, "Aquathermolysis of Heavy Oil Using Nano Oxides Of Metals", *J. Pet. Sci. Eng.*, 153, 385–390, https://doi.org/10.1016/j.petrol.2017.02.015

Le, J.-J., Wu, X.-L., Wang, R., Zhang, J.-Y., Bai, L.-L., Hou, Z.-W., 2015 "Progress in Pilot Testing of Microbial-Enhanced Oil Recovery in the Daqing Oilfield of North China", *Int. Biodeterioration & Biodegradation*, 97, January–February, 188–194. https://doi.org/10.1016/j.ibiod.2014.10.014

Li, H., Wang, Y., Qin, S., Cao, S., Yu, B., Sun, D., 2020, "Study on Viscosity Reduction Mechanism for Offshore Heavy Oil Under the Synergistic Action of Nanocatalyst and Ultrasound." Paper presented at the *Offshore Technology Conference Asia*, Kuala Lumpur, Malaysia, November. https://doi.org/10.4043/30398-MS

Li, Q., Liu, L., Sun, D., Xu, Z., 2024, "Recent Advances in Switchable Surfactants for Heavy Oil Production: A Review", *Energ. Geosci.*, 5, 100342–100356. https://doi.org/10.1016/j.engeos.2024.100342

Liang, C., Harjani, J. R., Robert, T., Rogel, E., Kuehne, D., Ovalles, C., Sampath, V., Jessop, P. G., 2012, "Use of CO2-Triggered Switchable Surfactants for the Stabilization of Oil-in-Water Emulsions," *Energy & Fuel*, 26, 488–494. https://doi.org/10.1021/ef200701g

Liu, L., Zhao, S., Zhou, B., Li, Z., Wang, Z., Wei, J., Pi, Y., Xing, M., Wang, J., 2025 "Evaluation of the Inhibitory Efficacy of Ionic Liquids on Asphaltene Precipitation in Crude Oil,"*J. Disp. Sci. Tech.*, June, 1–11. https://doi.org/10.1080/01932691.2025.2519393

Liu, Q., Qu, H., Liu, S., Zhang, Y., Zhang, S., Liu, J., Peng, B., Luo, D., 2020 "Modified Fe_3O_4 Nanoparticle Used for Stabilizing Foam Flooding for Enhanced Oil Recovery," *Coll. Surf. A: Physicochem. and Eng. Aspects*, 605, November. 20, 125383-125-395. https://doi.org/10.1016/j.colsurfa.2020.125383

Liu, Y., 2024, "The First Carbon Dioxide-Responsive Alkyamidine Surfactant Successful Production in China" *China Petroleum News*, 5 January. https://news.cnpc.com.cn/system/2024/01/05/030122187.shtml. Retrieved October 22, 2025.

Liu, Y., Jessop, P. G., Cunningham, M., Eckert, C. A., Liotta, C. L., 2006, "Switchable Surfactants," *Science*, 313, 958–960. https://doi.org/10.1126/science.1128142

Liu, Y., She, Y., Zhang, F., Feng, Q., Li, X., Dong, H., Sun, S., 2022, "Mechanism of Nano-depressurization and Injection-Augmenting Technology and Its Application in China's Oilfields: Recent Advances and Perspectives", *Energy Fuels*, 36 (18), 10751–10765. https://doi.org/10.1021/acs.energyfuels.2c01970

Mahdavi, E., Zebarjad, F. S., 2018, "Screening Criteria of Enhanced Oil Recovery Methods. Fundamentals of Enhanced Oil and Gas Recovery from Conventional and Unconventional Reservoirs", *Gulf Prof. Pub.*, 41–59. https://doi.org/10.1016/b978-0-12-813027-8.00002-3

Martinez-Palou, M. Likhanova, N. V., 2023 "Applications of Ionic Liquids in the Oil Industry: Towards a Sustainable Industry", *Bentham Sci. Pub. Pte. Ltd.* Singapore, ISBN: 978-981-5079-57-9.

Massarweh, O., Abushaikha, A. S., 2024, "Towards Environmentally Sustainable Oil Recovery: The Role of Sustainable Materials", *Energy Rep.* 12, 95–119.

Mierez, J., AlTammar, M. J., Alruwaili, K. M., Alfa, R. T., 2024 "Recent Advances of Ultrasound Applications in the Oil and Gas Industry", *Ultra. Sonochem.*, 103, 106767–106776. https://doi.org/10.1016/j.ultsonch.2024.106767

Moein-Elahi, S., Ahmadi-Koshoooei, M., Carbognani-Ortega, L., Scott, C. E., Chen, Z., Pereira-Almao, P., 2020, "Chemical Insight into Nano-Catalytic *In-Situ* Upgrading and Recovery of Heavy Oil", *Fuel*, 278, 118270–118279. https://doi.org/10.1016/j.fuel.2020.118270

Mohsenzadeh, A., Al-Wahaibi, Y., Jibril, A., Al-Hajri, R., Shuwa, S., 2015, "The Novel Use of Deep Eutectic Solvents for Enhancing Heavy Oil Recovery," *J. Pet. Sci. Eng.*, 130, 6–15, https://doi.org/10.1016/j.petrol.2015.03.018

Mullakaev, M. S., Abramov, V. O. Abramova, A. V., 2015, "Development of Ultrasonic Equipmentand Technology for Well Stimulation and Enhanced Oil Recovery", *J. Pet. Sci. Eng.*, 125, 201–208. https://doi.org/10.1016/j.petrol.2014.10.024

Mullakaev, M. S., Abramov, V. O. Abramova, A. V., 2017, "Ultrasonic Automated Oil Well Complex and Technology for Enhancing Marginal Well Productivity and Heavy Oil Recovery", *J. Pet. Sci. Eng*, 159, 1–7. https://doi.org/10.1016/j.petrol.2017.09.019

Nabipour, M., Ayatollahi, S., Keshavarz, P., 2017, "Application of Different Novel and Newly Designed Commercial Ionic Liquids and Surfactants for More Oil Recovery from an Iranian Oil Field," *J. Mol. Liq.*, 230, 579–588. https://doi.org/10.1016/j.molliq.2017.01.062

Naderi, K., Babadagli, T., 2007, "Clarifications on Oil/Heavy Oil Recovery Under Ultrasonic Radiation Through Core and 2D Visualization Experiments. " Presented at the *Petroleum Society's 8th Canadian International Petroleum Conference (58th Annual Technical Meeting)*, Calgary, Alberta, Canada, June 12–14. https://doi.org/10.2118/2007-032

Narehei, M. A., Shahri, M. P., Zamani, M., 2012, "Preparation and Characterization of Colloidal Gas Aphron based Drilling Fluids Using a Plant-Based Surfactant." Paper presented at the *SPE Saudi Arabia Section Technical Symposium and Exhibition*, Al-Khobar, Saudi Arabia, April. https://doi.org/10.2118/160888-MS

Nares, H. R., Schacht-Hernández, P., Ramírez-Garnica, M. A., Cabrera-Reyes, M. C., 2007 "Upgrading Heavy and Extraheavy Crude Oil with Ionic Liquid". Paper presented at the *International Oil Conference and Exhibition in Mexico*, Veracruz, Mexico, June. https://doi.org/10.2118/108676-MS

Nwosi-Anele, A. S., Obuebite, A. A., Okoro, E. E., Onyekonwu, M. 2019 "Heavy Oil Recovery: Environmental Implications and Mitigation Measures", *J. Oil, Gas and Coal Eng.*, 4(1), 34–38. https://doi.org/10.2118/172400-ms

Osma, L., García, L., Pérez, R., Barbosa, C., Botett, J., Sandoval, J., Manrique, E., 2019, "Benefit–Cost and Energy Efficiency Index to Support the Screening of Hybrid Cyclic Steam Stimulation Methods," *Energies*, 12, 4631. https://doi.org/10.3390/en12244631

Ovalles C. 2019, *Subsurface Upgrading of Heavy Crude Oils And Bitumen*, CRC Press, Boca Raton: Taylor & Francis Group, CRC Press; Chapter 3. https://doi.org/10.1201/9781315181028, and references therein.

Ovalles, C., Rivero, V., Salazar, A., 2015, "Downhole Upgrading of Orinoco Basin Extra-Heavy Crude Oil Using Hydrogen Donors under Steam Injection Conditions. Effect of the Presence of Iron Nanocatalysts," *Catalysts*, 5, 286–297. https://doi.org/10.3390/catal5010286

Pacific Northwest National Laboratory, 2025, "Nanomaterial- What are Nanomaterials?", *Nanomaterials* | PNNL. Retrieved August 26.

Pal, N., Saxena, N., Laxmi, K. V. D., Mandal, A., 2018, "Interfacial Behaviour, Wettability Alteration and Emulsification Characteristics of a Novel Surfactant: Implications for Enhanced Oil Recovery," *Chem. Eng. Sci.*, 187, 200–212. https://doi.org/10.1016/j.ces.2018.04.062

Pandey, L., Tiwari, P., 2022 "Microbial Enhanced Oil Recovery," Green Energy and Technology Series, Springer, https://doi.org/10.1007/978-981-16-5465-7

Patel, J., Borgohain, S., Kumar, M., Rangarajan, V., Somasundaran, P., Sen, R., 2015, "Recent Developments in Microbial Enhanced Oil Recovery," *Renew. Sust. Energy Rev.*, 52, 1539–1558. https://doi.org/10.1016/j.rser.2015.07.135

Paul, J. M., Davis, R. M., 1998, "Method for hydrotreating and upgrading heavy crude oil during production", US Patent No. 5,824,214.US5824214A

Petrobank Energy Resources, 2010, "Annual Progress Report (2009) for the Whites and Experimental Project", Revised May 14, IETP 01-019. Retrieved from Petrobank Annual Progress Report. Retrieved October 21, 2025.

Pourafshary, P., Farsi, H. A., 2019, "Enhancement of Heavy Oil Recovery by Nanoparticle/Microwave Application" *Eurasian Chemico-Technological J.*, 21, 51–56. https://doi.org/10.18321/ectj790

Prada, L., Botett, J., Contreras-Mateus, M. D., Hethnawi, A., Baakeem, S. S., Nassar, N. N., 2024, "Nanoparticles Technology for Improving Steam-Assisted Gravity Drainage Process Performance: A Review", *Ind. Eng. Chem. Res.*, 63, 13047−13077, https://doi.org/10.1021/acs.iecr.4c01377, and references therein.

Proton Technologies Canada, 2024 "Hydrogen Extraction Project", Final Technical Report Submission to Saskatchewan Petroleum Innovation Incentive, March 19. Hydrogen Extraction Project. Retrieved October 20, 2025.

Qi, L., Song, C., Wang, T., Li, Q., Hirasaki, G., Verduzco, R., 2018, "Polymer-Coated Nanoparticles for Reversible Emulsification and Recovery of Heavy Oil," *Langmuir*, 34(22), 6522–6528. https://doi.org/10.1021/acs.langmuir.8b00655

Quraishi, M., Bhatia, S. K., Pandit, S., Gupta, P. K. Rangarajan, V., Lahiri, D., Varjani, S., Mehariya, S., Yang, Y.-H., 2021, "Exploiting Microbes in the Petroleum Field: Analyzing the Credibility of Microbial Enhanced Oil Recovery (MEOR)", *Energies*, 14, 4684–4714. https://doi.org/10.3390/en14154684

Rodriguez, F., Delamaide, E., Rousseau, D., Bekri, S., 2022, "Which is the Most Attractive IOR Method to Produce the Venezuelan Highly Viscous Oil Resources in the Energy Transition Era? A Comprehensive Review of Research and Field Applications," Paper presented at the *ADIPEC*, Abu Dhabi, UAE, October. https://doi.org/10.2118/211344-MS

Saaid, I. M., Mahat, S. Q. A., Lal, B., Mutalib, M. I. A., Sabil, K. M.. 2014 "Experimental Investigation on the Effectiveness of 1-Butyl-3-methylimidazolium Perchlorate Ionic Liquid as a Reducing Agent for Heavy Oil Upgrading" *Ind. Eng. Chem. Res.*, 53 (19), 8279–8284. https://doi.org/10.1021/ie500499j

Safdel, M., Anbaz, M. M., Daryasafar, A., Jamialahmadi, M., 2017 "Microbial Enhanced Oil Recovery, a Critical Review on Worldwide Implemented Field Trials in Different Countries", *Renew. Sust. Ener. Rev.*, 74, July, 159–172. https://doi.org/10.1016/j.rser.2017.02.045

Saha, R., Uppaluri, R., Tiwari, P., 2019, "Impact of Natural Surfactant (Reetha), Polymer (Xanthan Gum), and Silica Nanoparticles to Enhance Heavy Crude Oil Recovery" *Energy & Fuels*, 33(5), 4225–4236. https://doi.org/10.1021/acs.energyfuels.9b00790

Sakthivel, S., Chhotaray, P. K., Velusamy, S., Gardas, R. L., Sangwai, J. S., 2015 "Synergistic Effect of Lactam, Ammonium and Hydroxyl Ammonium Based Ionic Liquids with and without NaCl on the Surface Phenomena of Crude Oil/Water System," *Fluid Phase Equil.*, 398, 80–97. https://doi.org/10.1016/j.fluid.2015.04.011

Salam, A. H., Nazar, M., Kamal, M. S., Ahad Mahboob, A., Hussain, S. M. S., Aljawad, M. S., Raza, A., Patil, S., 2025, "A Perspective on Ionic Liquids as Multifunctional Agents in Enhanced Oil Recovery and CO_2 Sequestration in Carbonate Formations,"*Carbon Capture Sci. Tech.*, 15, 100043–100052. https://doi.org/10.1016/j.ccst.2025.100443

Sawarkar, A. N., Pandit, A. B., Shriniwas, D. S., Joshi, J. B., 2009, "Use of Ultrasound in Petroleum Residue Upgradation", *Can. J. Chem. Eng.*, 87, 329–342. https://doi.org/10.1002/cjce.20169

Scheele-Ferreira, E., Scott, C. E., Perez-Zurita, M. J., Carbognani Ortega, L., Pereira-Almao, P. 2023 "Mo-Ni, Mo-Fe and Mo-Ni-W Nanoparticles for Down-Hole Upgrading", *Geoenergy Sci. Engineer*, 230, 212175–212185. https://doi.org/10.1016/j.geoen.2023.212175

Scott, C. E., Carbognani-Ortega, L., Pereira-Almao, P., 2019 "*In Situ* Upgrading via Hot Fluid and Nanocatalyst Injection", Chapter 6 in "*Advanced Catalytic Materials: Current Status and Future Progress*", Dominguez-Esquivel, J. M., Ramos, M. (eds), Springer, pp 129–149. ISBN: 978-3-030-25991-4. Advanced Catalytic Materials: Current Status and Future Progress | SpringerLink

Shaban, S., Dessouky, S., Badawi, A. E-F., Sabagh, A. E-S., Zahran, A., Mousa, M., 2014, "Upgrading and Viscosity Reduction of Heavy Oil by Catalytic Ionic Liquid", *Energy Fuels*, 28(10), 6545–6553. https://doi.org/10.1021/ef500993d

Shafiai, S. H., Gohari, A., 2020, "Conventional and Electrical EOR Review: The Development Trend of Ultrasonic Application in EOR", *J. Pet. Exp. Prod. Techn.*, 10, 2923–2945. https://doi.org/10.1007/s13202-020-00929-x

Shaikhah, D., Loise, V., Angelico, R., Porto, M., Calandra, P., Abe, A. A., Testa, F., Bartucca, C., Rossi, C. O., Caputo, P., 2024, "New Trends in Biosurfactants: From Renewable Origin to Green Enhanced Oil Recovery Applications," *Molecules*, 29, 301–336. https://doi.org/10.3390/molecules29020301

Sharifi, S., Behzadi, S., Laurent, S., Forrest, M. L., Stroevee, P., Mahmoudi, M., 2012 "Toxicity of Nanomaterials" *Chem. Soc. Rev.*, 41, 2323–2343. https://doi.org/10.1039/C1CS15188F

Singh, R., Panthi, K., Mohanty, K. K. 2017 "Microencapsulation of Acids by Nanoparticles for Acid Treatment of Shales," *Ener. Fuels*, 31(11), 11755–11764. https://doi.org/10.1021/acs.energyfuels.7b02003

Sircar, A., Rayavarapu, K., Bist, N., Yadav, K., Singh, S., 2022, "Applications of Nanoparticles in Enhanced Oil Recovery," *Pet. Res.*, 7 (1), 77–90. https://doi.org/10.1016/j.ptlrs.2021.08.004

Suarez-Suarez, R. G., Moein-Elahi, S., Orozco-Castillo, C. R., Chavez-Morales, S. M., Carbognani-Ortega, L. A., Scott-Algara, C. E., Pereira-Almao, P. R., 2023 "Heavy Oil Catalytic In Situ Upgrading in Fractured Carbonate Reservoirs. Fluid Distributions and Occurring Reactions Monitored with Marker Hydrocarbons", *Energy Fuels*, 37(21), 16503–16528.

Sun, S., Luo, Y., Zhou, Y., Xiao, M., Zhang, Z., Hou, J., Wei, X., Xu, Q., Sha, T., Dong, H., Song, H., Zhang, Z., 2017 "Application of Bacillus spp. in Pilot Test of Microbial Huff and Puff to Improve Heavy Oil Recovery", *Energy & Fuels*, 31(12), 13724–13732. https://doi.org/10.1021/acs.energyfuels.7b02517

Surguchev, L. M., Berenblyum, R. A., Surguchev, M. L., 2017 "Shift to Hydrogen - 100% Recovery from Depleted and Abandoned Gas Fields" *IOR 2017–19th European Symposium on Improved Oil Recovery*. April, Stavanger, Norway, 1–9. https://doi.org/10.3997/2214-4609.201700245

Suslick, K. S., Didenko, Y., Fang, M. M., Hyeon, T., Kolbeck, K. J., McNamara, W. B., Mdleleni, M. M., Wong, M., 1999, "Acoustic Cavitation and Its Chemical Consequences", *Phil. Trans. Roy. Soc.* A, 357, 335–353. https://doi.org/10.1098/rsta.1999.0330

Suwaid, M. A., Sitnov, S. A., Al-Muntaser, A., Yuan, C., Vakhin, A., Ancheyta, J., Varfolomeev, M. A., 2023 "Nanoparticles for Heavy Oil In Situ Upgrading", in "*Catalytic In-Situ Upgrading of Heavy and Extra-Heavy Crude Oils*," Chapter 6, Chichester, West Sussex, Wiley, 263–308. https://doi.org/10.1002/9781119871507.ch6

Temizel, C., Yegin, C., Hosgor, F. B., Aydin, H., Kabir, S., 2022, "Exploring Green Energy Applications in the Oil and Gas Industry." Paper presented at the *SPE Canadian Energy Technology Conference*, Calgary, Alberta, Canada, March. https://doi.org/10.2118/208900-MS

Thompson, L. H., Doraiswamy, L. K. 1999, "Sonochemistry: Science and Engineering", *Ind. Eng. Chem. Res.* 38, 1215–1249. https://doi.org/10.1016/j.ultsonch.2015.07.023

Turosung, S. N., Ghosh, B., 2017 "Application of Ionic Liquids in the Upstream oil Industry-A Review" *Int. J. Petrochem. Res.* 1(1), 50–60. https://madridge.org/international-journal-of-petrochemistry/ijpr-1000110.pdf

Wen, Q., Chen, J.-X., Tang, Y.-L., Wang, J., Yang, Z., 2015, "Assessing the Toxicity and Biodegradability of Deep Eutectic Solvents," *Chemosphere*, 132, 63–69, https://doi.org/10.1016/j.chemosphere.2015.02.061

Worthen, A., Taghavy, A., Aroonsri, A., Kim, I., Johnston, K., Huh, C., Bryant, S., DiCarlo, D., 2015, "Multi-Scale Evaluation of Nanoparticle-Stabilized CO_2-in-Water Foams: From the Benchtop to the Field." Paper presented at the *SPE Annual Technical Conference and Exhibition*, Houston, Texas, USA, September. https://doi.org/10.2118/175065-MS

Wu, T., Zhao, Y., Zhang, Y., Li, Z., & Su, J. 2023 "Matching Method Between Nanoparticle Displacement Agent Size and Pore Throat in Low Permeability Reservoir" *Front. Chem.*, 11, 1289271–1289288. https://doi.org/10.3389/fchem.2023.1289271

Xiong, R., Guo, J., Kiyingi, W., Feng, H., Sun, T., Yang, X., Li, Q., 2020 "Method for Judging the Stability of Asphaltenes in Crude Oil," *ACS Omega*, 5, 21420–21427. https://doi.org/10.1021/acsomega.0c01779

Xu, H. X., Pu, C. S., 2011, "Experimental Study of Heavy Oil Underground Aquathermolysis Using Catalyst and Ultrasonic," J. *Fuel Chem. Technol.*, 39, 606–610. https://doi.org/10.1016/S1872-5813(11)60037-6

Yan, K., Jie, W., Li, X., Horita, J., Stephens, J., Hu, J., Yuan, Q., 2023 "Microwave-Enhanced Methane Cracking for Clean Hydrogen Production in Shale Rocks" *Int. J. Hydrogen Energy*, 48(41), 15421–15432. https://doi.org/10.1016/j.ijhydene.2023.01.052

Yuan, C., Afanasev P., Askarova, A., Popov, E., Cheremisin, A., 2025, "*In situ* Hydrogen Generation in Subsurface Reservoirs of Fossil Fuels by Thermal Methods: Reaction Mechanisms, Kinetics, and Catalysis", *App. Ener*, 381, 125129. https://doi.org/10.1016/j.apenergy.2024.125219

Zabala, R., Franco, C. A., Cortés F. N., 2016, "Application of Nanofluids for Improving Oil Mobility in Heavy Oil and Extra-Heavy Oil: A Field Test," Paper presented at the *SPE Improved Oil Recovery Conference*, Tulsa, Oklahoma, USA, April. https://doi.org/10.2118/179677-MS

Zamani, A., 2010 "Transport of non-aqueous nano-dispersed catalyst suspensions in porous media." Doctoral thesis, University of Calgary, Calgary, Canada. Transport of non-aqueous nano-dispersed catalyst suspensions in porous media. Retrieved December 8, 2025

Zhang, J., Gao, H., Xue, Q. 2020, "Potential Applications of Microbial Enhanced Oil Recovery to Heavy Oil," *Crit. Rev. Biotechnol.* 40, 459–474, https://doi.org/10.1080/07388551.2020.1739618

5 Carbon Intensities and Future Trends

In this chapter, the carbon intensities (CIs) and greenhouse gas (GHG) emission reductions associated with various conventional and emerging technologies for Enhanced Oil Recovery (EOR) of heavy crude oils and bitumen (HCO/B) are discussed. The primary goal is to provide a high-level comparison of the technologies examined in Chapters 2 through 4, enabling readers to form their own assessments and conclusions. The upstream CIs are categorized into non-thermal and thermal processes, similar to the classifications presented in **Section 1.5** and illustrated in **Figures 1.4 and 1.5**, and the results are compared to the Base Cases (waterflooding for non-thermal processes and steamflooding for thermal technologies).

Next, future technological trends are presented, including hybrid EOR concepts as well as future Research and Development (R&D) directions for the next decade. It is essential to note that this monograph primarily discusses environmental concerns, including GHG carbon emissions, water usage, deforestation, and aerial footprint. Other ecological issues were not covered in detail, but they will be examined in the future. Those are air pollution, public health effects, noise and light pollution, and increased seismic activity via injection wells (Canbaz 2021).

5.1 UPSTREAM CARBON INTENSITY OF ECO-FRIENDLY TECHNOLOGIES

As mentioned in Chapter 1, several EOR technologies have been utilized with technical and economic success for the production of HCO and bitumen since the last century (Huc 2011, Ovalles 2019, Islam 2020, and Ahmadi 2023). Unfortunately, many of these processes are not energy efficient, generate a relatively large amount of contaminants (SO_x and NO_x) and GHGs, have high water consumption, and have a relatively large aerial footprint. As illustrated in **Figure 1.3**, the average CI for HCO is 80 kilograms of carbon dioxide equivalent removed for one US Barrel of petroleum (kg CO_2e/BBL), which is significantly higher than the emissions from light and medium crude oils, which average 40 kg CO_2e/BBL (Beckman 2018, Masnadi 2018, Lee 2021, and Carbon 2025). Higher values are reported for average HCO from California, Canada, Venezuela, and Indonesia, ranging from 148 to 190 kg CO_2e/BBL. Thus, HCO production has a greater environmental impact than medium and light crude oils.

As mentioned in Section 1.5, technology comparisons and evaluations can be a complex and challenging endeavor, as they depend on data availability and are subject to change as new information continuously emerges. Additionally, different stakeholders may have varying priorities and perspectives, resulting in less objective

DOI: 10.1201/9781003635475-5

assessments (Taber 1997 and Vishnumolakala 2020). Therefore, it is deemed essential to provide a high-level comparison of the most eco-friendly technologies. This comparison highlights viable technological pathways and emphasizes their CI, as measured in kg CO_2e/BBL.

5.1.1 Non-Thermal Routes

Table 5.1 presents the estimated upstream (CIs), the section in which was discussed in this monograph, and the Technology Readiness Level (TRL) for non-thermal EOR technologies for heavy crude oils and bitumen. This data account for emissions from exploration, drilling, completions, production, processing, and transportation. The routes are organized in alphabetical order following the Base Case, which is waterflooding, for reader's sake. The table also presents the percentage change of CI compared to the Base Case. A negative value indicates that the technological pathway results in a reduction of GHG emissions, while a positive value indicates an increase. Most of the upstream CIs in **Table 5.1** were taken from the literature (California 2023, Goavec 2023, Israel 2016, CHOPS 2025, Clean 2019, Masnadi 2018, Beckman 2018, Soiket 2019, Zhou 2019, Seidy-Esfahlan 2024, Arnold 2012, Sonoro 2024, Moradi 2023, Pinilla 2025, Wilson 2016, Farajzadeh 2022, Quintero 2024, Upreti 2007, and Lee 2021). They were calculated based on emissions from fuel combustion, electricity use, flaring and venting, as well as process-related activities such as water handling and artificial lift. Specific project values depend on field characteristics, field maturity, depth, thickness, petrographic properties, geographical location, facility design, and production duration. The estimates reported in **Table 5.1** do not include full lifecycle emissions such as transportation, refining, or combustion.

For the Base Case, **Table 5.1**, five crude oils currently undergoing waterflooding were utilized to determine the CI, as reported by CARB (California 2023). Those were Lost Hills (79.5 kg CO_2e/BBL) and Belridge South (104.7 kg CO_2e/BBL) from California, Peregrino (25.4 kg CO_2e/BBL) from Brazil, Napo (50.9 kg CO_2e/BBL) from Ecuador, and Maya (48 kg CO_2e/BBL) from Mexico. Based on the data, an average CI of 59 kg CO_2e/BBL was determined with a range in the 25–105 kg CO_2e/BBL (California 2023). Farajzadeh *et al.* (2022) reported that the CI of the crude oils produced by water injection strongly depends on the water utilization factor, that is, m^3 water/m^3 oil. For water cuts greater than 90%, which is typical of HCOs, the CI is expected to fall in the range of 30–60 kg CO_2e/BBL (Farajzadeh 2022). This range aligns closely with data provided by California Air Resources Board's (CARB's) data (California 2023). The latter was used for the Base Case, as shown in **Table 5.1**.

For the conventional HCO/B producing processes, the CIs were calculated by adding the emissions from diesel/gasoline combustion, electricity, sand, water, and waste logistics, and dividing by the barrels of oil produced, often over a year. According to the literature (see **Table 5.1**), the estimated GHG reductions for Mining without Upgrading (Case 3) (Israel 2016 and Alberta 2025), CHOPS (Case 4) (CHOPS 2002 and CHOPS 2025), and Cold Production (Case 6) (Masnadi 2018)

TABLE 5.1
Estimated Carbon Intensities (CIs) (Upstream Emissions Only), Section Discussed, and Technology Readiness Level for Non-Thermal Enhanced Oil Recovery (EOR) Processes of Heavy Crude Oils and Bitumen (HCO/B)

S. No.	Technology[a]	Section Discussed[b]	TRL[c]	Range kg CO_2e/BBL[d]	Average kg CO_2e/BBL[d]	% Change vs. Waterflooding[e]	Refereces
1	**Waterflooding (Base Case)**	–	**9**	**25–105**	**59**	–	**California (2023)**
2	ASP Flooding[f]	2.3.2	9	35–60	38	−36%	Goavec (2023)
3	Bitumen Mining without Upgrading	–	9	40–70	55	−7%	Israel (2016) Alberta (2025)
4	CHOPS[g]	–	9	40–60	50	−16%	CHOPS (2025)
5	CO_2 Injection	2.1.2	7–8	30–46	38	−36%	Clean (2019)
6	Cold Production	–	9	20–70	45	−24%	Masnadi (2018)
7	CSI-Biodegradable Solvents[h]	2.4.2	3	10–30	20	−66%	This work
8	Cyclic Solvent Injection	2.4.1	7–8	25–45	35	−41%	Beckman (2018) Soiket (2019)
9	Elect. Downhole Heaters	2.1.2	9	29–47	38	−36%	This work
10	Emulsion Injection	2.3.3	9	25–45	35	−41%	Zhou (2019) Seidy-Esfahlan (2024)
11	Enhanced CHOPS[g]	2.1.3	5	30–60	45	−24%	Arnold (2012) Sonoro (2024)
12	Ionic liquids	4.2	3	20–50	35	−41%	This work
13	Microbial Surfactants and Polymers (Microbial Enhanced Oil Recovery [MEOR])[i]	4.3.1	5	20–50	35	−41%	This work
14	Multilateral Wells and AICD[j]	2.1.1	9	10–30	20	−66%	Moradi (2023) Pinilla (2025)

(Continued)

TABLE 5.1 (Continued)

S. No.	Technology[a]	Section Discussed[b]	TRL[c]	Range kg CO_2e/BBL[d]	Average kg CO_2e/BBL[d]	% Change vs. Waterflooding[e]	Refereces
15	Nanomaterials	4.1	5	20–40	30	−50%	This work
16	Nitrogen Injection	2.2.2	9	20–50	35	−41%	This work
17	Nsolv™	2.5.3	6	13–39	26	−56%	Wilson (2016)
18	Plant-Based Surfactants	4.3.2	3	20–50	35	−41%	This work
19	Polymer Flooding	2.3.1	9	10–30 15–25	20	−66%	Farajzadeh (2022) Quintero (2024)
20	SSU-Cyclic C3-Injection[k]	2.5.2	5	30–45	38	−37%	This work
21	Switchable Surfactants	4.6	5	20–50	35	−41%	This work
22	VAPEX[l]	2.5.1	6	30–45	38	−37%	Upreti (2007) Lee (2021)

[a] Listed in alphabetical order after Waterflooding (Base Case). Estimated upstream carbon intensities based on emissions from fuel combustion, electricity use, flaring and venting, and process-related activities like water handling and artificial lift. Specific project values will depend on location, reservoir, and well productivity. Estimates do not include full lifecycle emissions such as transportation, refining, or combustion.
[b] Section number of the chapters in this book.
[c] TRL (see definition in **Table 1.2**).
[d] Typical average CI where BBL = US one barrel.
[e] Percentage of change of CI vs. waterflooding (Base Case). If negative means GHG emission reduction, whereas positive means GHG emission increase.
[f] ASP: Alkaline-Surfactant-Polymer.
[g] CHOPS: Cold Heavy Oil Production with Sand.
[h] CSI: Cyclic Solvent Injection.
[i] Microbial EOR.
[j] AICD: Autonomous Inflow Control Devices.
[k] SSU: Subsurface Upgrading.
[l] VAPEX: Vapor Extraction Process.

compared to waterflooding fall within a low range, specifically under 30% (−7, −16, and −24%, respectively). As highlighted in Section 1.5.1 and **Figure 1.4**, these technologies have a higher maturity (TRL = 9), are well-optimized, and show limited potential for further decreasing GHG emissions in the future.

In **Table 5.1**, three technologies focused on recent developments in Cold Production (Section 2.1) are reported. Those are Electrical Downhole Heaters (EDH, Case 9), Enhanced CHOPS (Case 14), and Multilateral Wells and Autonomous Inflow Control Devices (AICD) (Case 11). The CI for EDH was calculated using the field case data described in Section 2.1.2 and illustrated in **Figure 2.4**. The EDH had a rated power of 600 kW, and the test lasted for 96 hours. During this period, the oil rate increased from 980 to 1176 Barrels per day (BPD). As is known, the conversion from kg CO_2 emitted changes by region and energy mix (International 2024). Assuming a conversion factor range of 0.4–0.65 kg CO_2e/kWh, the CI for EDH was determined using equation 5.1. The results varied from 29.4 to 47.3 kg CO_2e/BBL. As shown in **Table 5.1**, an average value of 38 kg CO_2e/BBL with 36% reduction compared to waterflooding was determined.

$$CI_{\text{Electrical heaters}} = \frac{P_{\text{Electrical heaters}} \times T\ (\text{hours}) \times CF}{\left[(\text{Oil rate})_f - (\text{Oil rate})_o\right] \times T\ (\text{days})} \tag{5.1}$$

where

$CI_{\text{Elec.Heater}}$ = Carbon Intensity of Electrical Heater technology in kg CO_2e/BBL;
$P_{\text{Elec.Heaters}}$ = Rated power of the electrical heater (600 kW);
T = Time of the field test in hours or days (96 hours or 4 days);
CF = Conversion factor (from 0.4 to 0.65 kg CO_2e/kWh);
$(\text{Oil rate})_o$ = Oil rate before turning on the electrical heaters in BPD; and
$(\text{Oil rate})_f$ = Oil rate after turning on the electrical heaters in BPD;

For Enhanced CHOPS (Case 14, **Table 5.1**) using SuperSump system (Arnold 2012) and Sonoro Energy's multilateral well design (Sonoro 2024), the CI is estimated to be approximately 30–60 kg CO_2e/BBL, depending on well productivity, sand handling, and artificial lift efficiency. The average value (~45 kg CO_2e/BBL) is slightly lower than the baseline CHOPS (~50 kg CO_2e/BBL), with a 33% reduction compared to waterflooding (**Table 5.1**).

For Multilateral Wells and AICD (Case 14, **Table 5.1**), Moradi and coworkers reported a case study from an HCO field in the Marañon basin of the Peruvian Amazon, where reductions of the GHG footprint of 56–78% when AICD completions were utilized compared to cases without AICDs (Moradi 2023). Similarly, Pinilla *et al.* compared the performance of AICD systems, inflow control devices, and slotted liner completions for producing HCO from horizontal wells in the Llano Basin, Colombia (Pinilla 2025). These authors found decreases in CO_2 emissions of up to 82%, which was primarily attributed to a reduction in excessive water production (Pinilla 2025). Based on these values and as shown in **Table 5.1**, a CI range for Multilateral Wells and AICD was estimated as 10–30 kg CO_2e/BBL with an average value of 20 kg CO_2e/BBL (66% GHG reduction).

Table 5.1 presents the CIs for the immiscible carbon dioxide (Case 5) and nitrogen (Case 16) injection processes used in heavy oil EOR processes, which were previously discussed in Section 2.2. According to the literature, the CI for CO_2-based EOR is ~38 kg CO_2e/BBL (Clean 2019). This figure indicates a 36% reduction in CI compared to waterflooding. Conversely, currently, there is no data available comparing the reduction of GHG emissions from nitrogen injection to other EOR methods. One clear advantage is that no GHGs will be back-produced as the cases for CO_2 and CH_4 flooding. The only GHG emissions to consider will come from the air separation unit involved in the process. Based on the energy consumed reported in the literature, the CI ranges for Pressure Swing Adsorption and Cryogenic separation are estimated to be 20–30 kg CO_2e/BBL and 30–45 kg CO_2e/BBL, respectively (Tileuberdi, 2024). For remote fields powered by diesel fuel, the CI is slightly higher, 40–50 kg CO_2e/BBL. As indicated in **Table 5.1**, Nitrogen Injection EOR recovery of HCO (Case 16) has an estimated CI range of 20–50 kg CO_2e/BBL, with an average of 35 kg CO_2e/BBL. This value represents a 41% reduction compared to waterflooding.

Table 5.1 outlines three technologies related to Chemical EOR (Section 2.3). Those are ASP Flooding (Case 2), Emulsion Injection (Case 10), and Polymer Flooding (Case 19). For ASP Flooding, the synthesis and transport of the alkali, surfactant, and polymer, as well as the high injection volumes, are the most energy-consuming steps (Goavec 2023 and Chakraborty 2025). Based on this information, the CI can be estimated to range from 35 to 60 kg CO_2e/BBL, with an average of approximately 38 kg CO_2e/BBL, representing a 37% reduction compared to waterflooding.

There is limited information available in the literature regarding the CI of Emulsion Injection (Case 10, **Table 5.1**) used for HCO recovery. It can be speculated that the main contributors to CO_2 emission include emulsion preparation, surfactant synthesis and use, and fluid downhole injection. Based on the field tests conducted (Zhou 2019 and Seidy-Esfahlan 2024), the estimated CI ranges from 25 to 45 kg CO_2e/BBL, with an average of 35 kg CO_2e/BBL. This represents a reduction of 41% compared to waterflooding.

In the case of Polymer Flooding (Case 19, **Table 5.1**), Farajzadeh *et al.* (2022) reported that once the injection of polymer reduces the water cut, CO_2 emissions are also reduced for a field in the Middle East. The authors found a reduction in the CO_2 intensity from ~55 to ~20 kg CO_2/BBL. Similarly, Quintero and coworkers observed a slight decrease in CI compared to the baseline, from 20 kg CO_2e/BBL to 19.3 kg CO_2e/BBL, in the Chichimene field (Colombia) (Quintero 2024). Reports indicate that Polymer Flooding can help meet environmental targets by reducing CO_2 emissions per barrel by three to six times (Dupuis 2020). Based on these findings, a medium-to-high average CI of 20 kg CO_2e/BBL is achieved, representing a 66% reduction compared to the Base Case (see **Table 5.1**).

Two Solvent-Only Injection technologies (Section 2.4) are presented in **Table 5.1**. Those are Cyclic Solvent Injection (CSI) using Biodegradable Solvents (Case 7) and Hydrocarbon Solvents (Case 8). The CI of these processes depends strongly on the solvent-to-oil ratio used and the amount of diluent lost in the reservoir vs. that which is recycled. Imperial Oil utilized a proprietary flow-assurance solvent to minimize the formation of heavy liquids during a CSI pilot test (Beckman 2018). This author reported a CI of 24 kg CO_2e/BBL (Beckman 2018). Similarly, Soiket *et al.* investigated the GHG emissions in a propane-based extraction process. These authors estimated a range from 24.8 to 29.1 kg CO_2e/BBL (Soiket 2019). Based on these findings and assuming a conservative 70% solvent recovery, the estimated CI range for CSI using Hydrocarbon Solvents (Case 8, **Table 5.1**) is 25–45 kg CO_2e/BBL, with a medium average of 35 kg CO_2e/BBL. This value represents a 41% reduction compared to waterflooding.

For CSI using Biodegradable Solvents (Case 7, **Table 5.1**), further reductions in CIs can be envisioned compared to CSI utilizing hydrocarbon solvents. As these "green" solvents are plant-based, their CI is speculated to be in the 10–30 kg CO_2e/BBL range, with an average of 20 kg CO_2e/BBL. This value represents a 66% reduction compared to waterflooding.

Table 5.1 registers three Subsurface Solvent Deasphalting routes (Section 2.5). Those are Nsolv™ (Case 17), Cyclic C3-Injection (Case 20), and VAPEX (Case 21). For Case 17, the report from Alberta Innovates estimated a range between 13 and 39 kg CO_2e/BBL, with an average of 26 kg CO_2e/BBL (Wilson 2016). This value translates into a 56% reduction compared to waterflooding.

In Case 22 of **Table 5.1**, the literature indicates that the implementation of VAPEX can reduce GHG emissions by 70–80% compared to Steam-Assisted Gravity Drainage (SAGD) (Upreti 2007 and Lee 2021). This reduction is attributed to lower energy consumption for steam generation and varies based on location and reservoir characteristics. With a CI of 151 kg CO_2e per barrel (see Section 5.2) for SAGD, we can estimate the CI range for VAPEX to be between 30 and 45 kg CO_2e/BBL. Consequently, the average value of 38 kg CO_2e/BBL reflects a 37% reduction compared to waterflooding. For Cyclic C3-Injection (Case 20, **Table 5.1**), no new development of this technology has been reported until the publication of this book, and no estimation of its CI has been documented in the literature. Similar to VAPEX, cyclic propane injection eliminates the need for steam to mobilize HCO. Thus, the same CI values can be estimated as presented in **Table 5.1**. However, the limited number of field trials of VAPEX and Cyclic C3-injection adds more technical uncertainty over the GHG intensity and oil recovery factor compared to more mature processes.

In the area of Emerging Technologies (Chapter 4), five non-thermal routes are examined in **Table 5.1** for the EOR of heavy crude oils and bitumen. Those are Ionic liquids (IL, Case 12), Microbial Surfactants and Polymers (MEOR, Case 13), Nanomaterials (Case 15), Plant-Based Surfactants (Case 18), and Switchable Surfactants (Case 21). There are limited studies on the CI of these technological routes. However, by comparing them to other environmentally friendly technologies, estimated CI values can be inferred. For IL (Case 12), the primary GHG emissions arise from the synthesis, transportation, handling, and downhole injection of these materials. Moreover, the potential reduction in GHG emissions can be attributed to the ability of ILs to enhance the efficiency of HCO recovery, while decreasing reliance on energy-intensive conventional production and upgrading processes (Martinez-Palou 2023, Salam 2025). As a result, a conservative assessment indicates a medium potential for reducing GHG emissions of 20–50 kg CO_2e/BBL. The average value of 35 kg CO_2e/BBL reflects a 37% reduction compared to waterflooding.

For nanofluids, nanoemulsions, and nanocatalysts (Case 15, **Table 5.1**), the main GHG emission steps are the preparation, transportation, and underground injection. Since these materials are fluids, the same procedures are used for synthesizing and handling oilfield chemicals (e.g., demulsifiers, antifoaming agents) and solid suspensions (e.g., drilling muds). These technologies have been field-tested in non-thermal (waterflooding) conditions and have demonstrated their effectiveness in removing asphaltenes, reducing viscosity, and restoring wettability (Zabala 2016, Wu 2023, Cárdenas 2024, Franco 2025). Based on this information, a conservative assessment suggests a medium potential for reducing GHG emissions of 20–40 kg CO_2e/BBL. The average value of 50 kg CO_2e/BBL reflects a 37% reduction compared to traditional waterflooding.

In the cases of MEOR (Case 13), Plant-Based Surfactants (Case 18), and Switchable Surfactants (Case 21), the primary GHG emissions are generated during the extraction, purification, formulation, and transportation of biomaterials. Factors such as CAPEX, OPEX, the cost of bio-additives and nutrients, production scale, and market variability significantly influence CI (Quraishi 2021 and Hama 2023). A conservative estimate suggests a moderate potential to reduce GHG emissions by 20–50 kg CO_2e/BBL. The average reduction of 35 kg CO_2e/BBL represents a 37% decrease compared to waterflooding.

The results in **Table 5.1**, consistent with **Figure 1.4**, show that the environmentally friendly technologies with the highest percentage reduction in GHG emissions compared to the Base Case (waterflooding) are as follows: CSI-Biodegradable Solvents (Case 7), Multilateral Wells amd AICD (Case 14), Nanomaterials (Case 15), and Polymer Flooding (Case 19). However, it is important to note that **Table 5.1** provides only a high-level comparison of the most eco-friendly technologies. Some processes are well-studied, producing large volumes of oil, whereas others are based on limited data and apply to only a small portion of the resource base. Furthermore, each project should be analyzed individually, taking into account factors such as location, reservoir characteristics, availability of raw materials, facility design, production duration, and the company's expertise, among other relevant considerations.

5.1.2 Thermal Processes

As discussed in Chapter 3, thermal methods for heavy crude oil production involve applying heat to lower the oil's viscosity, making it easier to extract from reservoirs where non-thermal technologies have either been exhausted, proven ineffective, or are economically unfeasible. Generally, thermal processes can enhance recovery factors by up to 40–70% of the OOIP.

Table 5.2 offers the estimated upstream CIs, the section discussed, and the TRLs for thermal EOR technologies for HCO/B. As previously mentioned, most of the upstream CIs were taken from the literature. These estimates were calculated based on emissions from fuel combustion, steam generation, electricity use, flaring and venting, as well as process-related activities such as water handling and artificial lift. Specific project values depend on field characteristics, field maturity, depth, thickness, petrographic properties, geographical location, facility design, and production duration. The values reported in **Table 5.2** do <u>not</u> include transportation, refining, or

TABLE 5.2
Estimated Carbon Intensities (CIs) (Upstream Emissions Only), Section Discussed, and Technology Readiness Level (TRL) for Thermal Enhanced Oil Recovery (EOR) Processes of Heavy Crude Oils and Bitumen (HCO/B)

S. No.	Technology[a]	Section Discussed[b]	TRL[c]	Range kg CO_2e/BBL[d]	Average kg CO_2e/BBL[d]	% Change vs. Steamflooding[e]	Ref.
1	**Steamflooding (Base Case)**	**–**	**9**	**53–98 92–179 160–180**	**117**	–	**This work California (2023) Gordon (2015)**
2	Acceleware RF XL[f]	3.3.2	6	16–22	19	−84%	Okoniewski (2025)
3	CSS-Solvent Co-injection[g]	3.1.1	9	40–120 40–70	80	−32%	Leaute (2007), Lee (2021)
4	Cyclic Steam Stimulation (CCS)	–	9	61–109 106.7 80–120 60–110 106	94	−20%	Orellana (2018) Beckman 2018, Soiket (2019), Pérez (2022), Clean (2023)
5	eMSAGP[h]	3.2.2	7–8	56 60	58	−50%	Chen (2019), Sleep (2021)
6	eMVAPEX[i]	3.2.3	7–8	40–55	48	−59%	MEG (2024)
7	ESEIEH[j]	3.3.3	6	44 20–35 10–88	40	−66%	Oni (2023), Morris 2021, Safaei (2019)
8	Geothermal/Steam	3.4.3	5	11–17	14	−88%	Cano (2022)
9	Nanomaterials	4.1	5	32–62	49	−58%	Franco (2025)
10	Nuclear Energy EOR/Steam[k]	3.4.2	4	3–9	6	−95%	Badruzzaman (2008)
11	SAGD[l]	–	9	49–102 73–112 68–75 94 66	81 (57)[m]	−31% (−51%)[m]	Orellana (2018), Oni (2023), Morris (2021), Chen (2019), Sleep (2021)

(Continued)

TABLE 5.2 (Continued)

S. No.	Technology[a]	Section Discussed[b]	TRL[c]	Range kg CO_2e/BBL[d]	Average kg CO_2e/BBL[d]	% Change vs. Steamflooding[e]	Ref.
12	Solar Energy EOR/Steam[n]	3.4.1	7–8	21–72	47	−60%	Sandler (2012)
13	 Solvent-SAGD Co-Injection	3.1.3	9	38–86 46–86	62	−44%	Chen (2019), Hoque (2023)
14	Steamflooding-solvent	3.1.2	9	71–136	104	−11%	Castellanos-Diaz (2016)
15	Steam/Additive Co-Injection	3.1.4	9	30–60[o] 40–90[p]	60	−49%	Pérez (2022)[o], Sanchez (2022)[p]
16	Steam/NCG Co-Injection[q]	3.2.1 3.2.4	7–8	60–65	63	−46%	ConocoPhillips (2017)
17	Ultrasound Energy	4.4	6	20–60	40	−66%	This work

[a] Listed in alphabetical order after Steamflooding (Base Case). Estimated upstream carbon intensities based on emissions from fuel combustion, electricity use, flaring and venting, and process-related activities like water handling and artificial lift. Specific project values will depend on location, reservoir, and well productivity. Estimates do not include full lifecycle emissions such as transportation, refining, or combustion.
[b] Section number of this book in which the technology was discussed.
[c] TRL (see definition in **Table 1.2**).
[d] Typical average intensity where BBL = US one barrel.
[e] Percentage of change of CI vs. steamflooding (Base Case). If negative means GHG emission reduction, whereas positive means GHG emission increase.
[f] Electromagnetic heating using Acceleware RF XL technology.
[g] CSS: Cyclic Steam Stimulation.
[h] Enhanced Modified Steam and Gas Push.
[i] Enhanced Modified Vapor Extraction.
[j] Electromagnetic heating using Enhanced Solvent Extraction Incorporating Electromagnetic Heating (ESEIEH) process.
[k] Only nuclear-generated steam. Other operational cost not included.
[l] Solvent-Assisted Gravity Drainage.
[m] Recent values (not peer reviewed) (Canadian 2025).
[n] Solar steam is utilized for one-third for the total steam generation.
[o] Steam/Foam co-injection.
[p] Steam/surfactant co-injection.
[q] NCG: Non-Condensable Gases (natural gas and flue gas).

combustion processes. The routes are organized in alphabetical order, starting with the Base Case: steamflooding. The table also presents the percentage change of CI compared to the Base Case. A negative value indicates that the technological pathway results in a reduction of GHG emissions, while a positive value indicates an increase.

For steamflooding (Base Case, **Table 5.2**), a relatively straightforward calculation allows us to estimate the minimum CI of this EOR process. If we assume a steam-to-oil ratio (SOR) of 2.5, the energy required to generate saturated steam from 25°C to 190°C at 10–15 bar with an efficiency (η) of 85% is 1.35 ± 0.4 GJ/BBL of oil as illustrate in equations 5.2–5.5.

Mass of steam (m) in kg per BBL oil:

$$m = \text{SOR} \cdot \left(0.159\,\text{m}^3/\text{BBL}\right) \times \cdot \left(1{,}000\,\text{kg}/\text{m}^3\right) = 2.5 \times 159 \approx 397.5\,\text{kg}/\text{BBL} \tag{5.2}$$

Energy (q) to raise feedwater to saturated steam (per kg):

$$q \approx c_p\left(T_{sat} - T_{fw}\right) + h_{vap} \tag{5.3}$$

where c_p is the average water heat capacity (≈4.27 kJ/kg C) and h_{vap} is the average heat of vaporization of water (≈2,196 kJ/kg at 190°C) in the temperature range from $T_i = 25^\circ\text{C}$ to $T_{sat} = 190^\circ$ (Water 2025).

$$q \approx 4.27 \cdot \times (190 - 25) + 2{,}196 = 2{,}900.6\,\text{kJ}/\text{kg} \tag{5.4}$$

Fuel energy (E_{fuel}) with Boiler efficiency (η):

$$E_{fuel} = (m \cdot \times q)/\eta = (397.5\text{ kg/BBL} \times 2{,}900.6\text{ kJ/kg})/0.85 = 1.35\,\text{GJ/BBL} \tag{5.5}$$

Where $\eta = 0.85$ (Boiler 2025).

Depending on steam pressure, initial water temperature, and boiler efficiency, the energy required to generate saturated steam can range from 0.95 to 1.75 GJ/BBL. For the combustion of natural gas, the conversion factor is 56 kg CO_2e/GJ (Methodology 2006). By using equation 5.6, the CI of steamflooding varied from 53 kg CO_2e/BBL (calculated as 0.95 GJ/BBL × 56 kg CO_2e/GJ) to 98 kg CO_2e/BBL (calculated as 1.75 GJ/BBL × 56 kg CO_2e/GJ).

$$\text{CI (kg CO}_2\text{e/BBL)} = E_{fuel}\text{ (GJ/BBL)} \times 56\text{ kg CO}_2\text{e/GJ} \tag{5.6}$$

However, it is essential to note that these values account only for steam generation; they do not include other factors such as electricity use, flaring and venting, as well as process-related activities such as water handling and artificial lift. Some operators have reported SOR as low as 2. Most steamflooding projects typically operate within a range of 2.5–4, depending on the reservoir characteristics and oil quality. Therefore, this calculation reflects the minimum CO_2 emissions that steamflooding would produce with an SOR of 2.5.

Conversely, the maximum CI values were estimated using the reported CI for seven mature California oil fields currently undergoing steamflooding, with SORs in the 6–10 range (California 2023). The oilfields and their respective CIs are as follows: Midway-Sunset (179.5 kg CO_2e/BBL), Cymric (95.9 kg CO_2e/BBL), San Ardo (161.8 kg CO_2e/BBL), Coalinga (158 kg CO_2e/BBL), McKittrick (155.1 kg CO_2e/BBL), and Kern River (92.4 kg CO_2e/BBL). From this data, California's CI ranges from 92 to 179 kg CO_2e/BBL (California, 2023). Additionally, the Duri oil field in Indonesia has been undergoing continuous steam injection for over 40 years and also has a high SOR (Steamflood 2018). The estimated upstream GHG emissions were reported to range from 160 to 180 kg CO_2e/BBL (Gordon 2015). If we consider the straightforward calculation and the California and Duri data, the CI range for steamflooding can be established as 53–180 kg CO_2e/BBL. Thus, it can be assumed that the overall average CI is 117 kg CO_2e/BBL, which is used for the Base Case, as shown in **Table 5.2**. As expected, this data highlight that steamflooding operations are generally more carbon-intensive than, for example, waterflooding. As shown in **Table 5.1**, the latter averages around 59 kg CO_2e/BBL.

Two of the most conventionally utilized thermal routes are cyclic steam injection (CSS, Case 4, **Table 5.2**) and (SAGD, Case 11, **Table 5.2**). These technologies have been successfully implemented for the EOR of HCO/B and bitumen with multi-well and multi-field applications worldwide (TRL 9). Various authors have estimated the CI of CSS for different oil fields around the globe (Orellana 2018, Beckman 2018, Soiket 2019, Pérez 2022, and Clean 2023). **Table 5.2** shows a range from 61 to 120 kg CO_2e/BBL, with an average of 94 kg CO_2e/BBL. This value represents a 20% reduction in GHG emissions compared to the Base Case.

For SAGD (Case 11, **Table 5.2**), several research groups have estimated the carbon efficiency since 2018 (Orellana 2018, Oni 2023, Morri 2021, Chen 2019, and Sleep 2021). The resulting energy consumption and GHG emissions vary depending on steam and electrical energy demand, operating conditions such as boiler efficiency and heat integration, and the composition of bitumen, produced gases, and water recovered (Soiket 2019). The reports indicate that CI can range between 49 and 112 kg CO_2e/BBL, with an average of 81 kg CO_2e/BBL, representing a 31% reduction compared to steamflooding (see Base Case, **Table 5.2**). However, recent technological optimization, such as methane co-injection, electricity co-generation, and AICDs, has led to a further decrease in the CI of SAGD to 57 kg CO_2e/BBL (Hannouf 2022 and Canadian 2025). These improvements represent a 51% decrease in GHG emissions compared to the Base Case (referenced in brackets, Case 11, **Table 5.2**).

For the Low-Carbon Intensity Steam Injection Technologies (Section 3.1), **Table 5.2** outlines several cases: CSS-Solvent Co-injection (Case 3), Solvent-SAGD Co-injection (Case 13), Steamflooding-Solvent (Case 14), and Steam/Additive Co-Injection (Case 15). For the first technology, Lee and Babadagli (2021) conducted a comparative analysis of GHG emissions from steam/solvent processes relative to steam-based technologies. These authors found that the CI for CSS-Solvent Co-Injection varied from 40 to 70 kg CO_2e/BBL (Lee 2021). Additionally, Leaute (2007) reported 25% GHG emission reduction by using Imperial Oil's Liquid Addition to Steam for Enhancing Recovery (LASER). Thus, the CI range for CSS-Solvent Co-injection (Case 3) was established as 40–120 CO_2e/BBL with an average of

80 CO_2e/BBL. This value represents a 32% reduction compared to steamflooding (**Table 5.1**, Case 1).

As discussed in Section 3.1.2, Castellanos-Diaz and colleagues from Shell reported a 32% reduction in SOR compared with the no-solvent case during the field test of the Vertical-well Steam Drive process in Peace River, Canada (Castellanos-Diaz 2016). Using this value, the CI range for Steamflooding-Solvent (Case 14, **Table 5.2**) is estimated to range from 36 to 122 kg CO_2e/BBL, with an average of 79 kg CO_2e/BBL.

Solvent-SAGD Co-injection (Case 13, **Table 5.2**) is also known as Expanded Solvent SAGD (ES-SAGD) or Solvent-Assisted SAGD (SA-SAGD) because the solvent is co-injected with steam in a vapor phase, as discussed in Section 3.1.3. Chen and coworkers modeled the environmental and economic benefits of SA-SAGD (Chen 2019). The numerical results indicated that, with a liquid-pool level of 3 m, GHG emissions would be reduced by 30% with C3/C5 solvents and by 16.5% with C7 solvents, compared to traditional SAGD methods (Chen 2019). Hoque conducted a study of the GHG emission footprint of solvent-steam-based bitumen extraction (Hoque 2023). The findings revealed an emission range of 49.7–82.6 kg CO_2eq/BBL at a 90% confidence level (Hoque 2023). Using this data, the CI range for Solvent-SAGD Co-injection (Case 13, **Table 5.2**) is estimated to be from 38 to 86 kg CO_2e/BBL, with an average of 62 kg CO_2e/BBL. This value represents a 47% decrease in emissions compared with steamflooding.

For Steam/Additive Co-injection (Case 15, **Table 5.2**), the CIs were sourced from the literature. Pérez *et al.* (2022) determined the CI for the steam-foam co-injection and compared it with the steam-only case. Their findings indicated that the CI for the CSS/foam co-injection was lower, at approximately 45 kg CO_2e/BBL, compared to 94 kg CO_2e/BBL for conventional CSS. Additionally, Sanchez from ConocoPhillips (**Table 3.2**, entry 8) used steam/surfactant co-injection to improve SAGD performance in a field pilot (Sanchez 2022). The results of this study demonstrated a 20% reduction in GHG emissions intensity compared to conventional SAGD (Sanchez 2022). Based on this data, the estimated CI range for Steam/Additive Co-injection (Case 15, **Table 5.2**) is from 30 to 90 kg CO_2e/BBL, with an average of 60 kg CO_2e/BBL. This value represents a 47% decrease in emissions compared with steamflooding.

In the area of Steam and Non-Condensable Gases Co-Injection (Section 3.2), three technologies for the EOR of HCO/B are included in **Table 5.2**. Those are: Enhanced Modified Steam and Gas Push (eMSAGP, Case 5), Enhanced Modified Vapor Extraction (eMVAPEX, Case 6), and Steam/Non-Condensable Gases Co-Injection (NGC, Case 16). The latter includes natural gas (Section 3.2.1) and flue gas (Section 3.2.4) routes. For the eMSAGP process (Case 5, **Table 5.2**), Chen et al. (2019), and Sleep *et al.* (2021) conducted life-cycle analyses of the commercial application of this technology and). The findings indicated a CI range of 56–60 kg CO_2e/BBL with an average of 58 kg CO_2e/BBL. The value represents a 50% decrease vs. steamflooding. For eMVAPEX (Case 6, **Table 5.2**), MEG Energy Corp reported a 43% reduction of GHG emissions compared with conventional SAGD (MEG 2024). This optimization translates to a CI range of 40–55 kg CO_2e/BBL, with an average of 58 kg CO_2e/BBL, representing a 59% reduction compared to steamflooding.

For the Steam/NGC co-injection (Case 16, **Table 5.2**), ConocoPhillips reported that GHG emission reduction is 20–25% compared to SAGD in a 2017 field project in Canada (ConocoPhillips 2017). Based on this information, the CI of steam/flue gas co-injection (Case 16) has a CI range of 60–65 kg CO_2e/BBL, with an average of 63 kg CO_2e/BBL, representing a 46% reduction compared to conventional steamflooding.

Two electromagnetic heating technologies (Section 3.3) were included in **Table 5.2**. Those are: Acceleware RF XL (Case 2, Section 3.2.2) and Enhanced Solvent Extraction Incorporating Electromagnetic Heating (ESEIEH, Case 7) process (Section 3.3.3). For the first technological route, a third-party Life Cycle Analysis (LCA) yielded a CI range of 16–22 kg CO_2e/BBL, with an average of 19 kg CO_2e/BBL (Okoniewski 2025). This value represents a significant 84% reduction in carbon emissions compared to traditional steamflooding methods. For ESEIEH (Case 7, **Table 5.2**), the range and the average CI were estimated at 20–60 kg CO_2e/BBL and 40 kg CO_2e/BBL, respectively. The latter represents a 66% reduction compared to the Base Case.

In the topic of Use of Low-Carbon Intensity Energy (Section 3.4), three technologies for the EOR of HCO/B are included in **Table 5.2**. Those are: Geothermal/Steam (Case 8, 3.4.3), Nuclear Energy EOR/Steam (Case 10, Section 3.4.2), and Solar Energy EOR/Steam (Case 12, Section 3.4.1). Cano and coworkers reported that, on average, conventional geothermal plants (Case 8, **Table 5.2**) produce ~14 ± 3 kg CO_2/BBL (Cano 2022). Therefore, a CI range of 11–17 kg CO_2e/BBL, with an average of 14 kg CO_2e/BBL. This value represents an 88% reduction compared to conventional steamflooding methods.

Badruzzaman and colleagues calculated the annual CO_2 reduction per barrel for heavy crude oil and bitumen production in California and Alberta, Canada, utilizing nuclear-generated steam (Case 10, **Table 5.2**) (Badruzzaman 2008). Assuming a steam-to-oil ratio of 3.32, the authors found that using nuclear-generated steam resulted in a significant reduction (~90%) in CO_2 emissions compared to natural gas. Similarly, Maio and coworkers reported that integrating a high-temperature gas-cooled reactor (HTGR) with the SAGD process led to almost eliminating CO_2 emissions from the process. However, CO_2 emissions from imported power remain (Maio 2010). Therefore, a CI range of 3–9 kg CO_2e/BBL, with an average of 6 kg CO_2e/BBL. This value represents a 95% reduction compared to conventional steamflooding.

For Solar Energy EOR/Steam (Case 12, **Table 5.2**), Sander *et al.* assumed that solar steam would be utilized for one-third of the time. To ensure consistent steam injection rates, this system is paired with natural gas generation at night. As a result, there is a 60% reduction in emissions compared to conventional steamflooding methods. This reduction translates to 47 kg of CO_2e/BBL for Solar Energy EOR/Steam (Case 12, **Table 5.2**).

Two emerging technologies are reported in **Table 5.2**. Those are Nanomaterials (Case 9) and Ultrasound Energy (Case 17). Franco *et al.* (2025) conducted a field test in a Colombian heavy oil reservoir under cyclic steam stimulation. They found a 48% reduction in operational costs, mainly due to decreased steam generation and lower GHG emissions compared to CSS (Case 4, **Table 5.2**). Thus, an average value of 49 kg of CO_2e/BBL can be estimated, corresponding to a 58% decrease compared to steamflooding.

For the Ultrasound Energy technology (Case 17, **Table 5.2**), low energy consumption and small environmental impact have been reported by various authors, suggesting that this technology has a medium-to-high potential to reduce GHG emissions (see **Figure 1.5**) (Sawarkar 2009, Shafiai 2020, Hamidi 2021, Agi 2022, Mierez 2024, and Gomaa 2024). However, due to the relatively low penetration of ultrasonic energy, this technology is less effective for deeper reservoir applications, particularly beyond 10 m. In comparison with other thermal processes and depending on the electricity cost, it can be speculated that the CI is around 20–60 kg CO_2e/BBL.

The results in **Table 5.2**, consistent with **Figure 1.5**, show that the environmentally friendly technologies with the highest percentage reduction in GHG emissions compared to the Base Case (steamflooding) are as follows: Acceleware Radiofrequency (RF) heating (Case 2), eMVAPEX (Case 6), ESEIEH (Case 7), Geothermal/Steam (Case 8), Nuclear Energy EOR/Steam (Case 10), Solar Energy EOR/Steam (Case 12), Solvent-SAGD Co-injection (Case 13), and Ultrasound Energy (Case 17). As anticipated, a comparison of the results between **Tables 5.1** and **5.2** reveals that the thermal technologies outlined in the latter generally have higher CIs than those listed in the former table (non-thermal). Similar findings have been reported in the literature by Lee and Babadagli (Lee 2021).

As mentioned earlier, it is essential to clarify that **Tables 5.1** and **5.2** provide only a high-level comparison of the most environmentally benign technologies. Some processes are well-established, generating significant revenues, whereas others are based on a small number of pilots and lab experiments. Furthermore, each project should be analyzed individually, considering factors such as location, reservoir characteristics, production duration, and the company's expertise, among other relevant considerations.

5.2 HYBRID CONCEPTS

Over the last decade, hybrid concepts for EOR of HCO/B have been proposed in the literature. These concepts integrate several eco-friendly technologies to maximize oil recovery while minimizing environmental impact. For example, Ashrafi *et al.* showed that SAGD facilities, when combined with post-combustion carbon capture technology, can improve efficiency by 13–47%, depending on the specific configuration of the plant (Ashrafi 2021). This integration can lead to reduced energy consumption and GHG emissions, resulting in a significant enhancement in boiler and combustion performances (Ashrafi 2021).

Alsabagh et al. (2023) proposed that integrating steam injection with polymers, biopolymers, and surfactants can enhance oil recovery by reducing interfacial tension between oil and water while increasing the viscosity of the injected liquid. Additionally, Dong and coworkers examined hybrid thermal-non-condensable gases, hybrid thermo-solvent, and hybrid thermochemical processes to overcome the limitations of conventional steam injection techniques and reduce CIs (Dong 2021).

Shaikhah *et al.* (2024) explored the synergistic effects of biosurfactants with other EOR agents, such as polymers, nanoparticles, gases, or enzymes, to improve oil recovery efficiency, reduce operational costs, and mitigate environmental impacts.

These authors suggested that developing reliable, robust models and methods to predict and optimize biosurfactant performance and efficiency in the reservoir will be of paramount importance in the future.

Ovalles et al. (2017) numerically evaluated the integration of one well-pair SAGD with solvent deasphalting (SDA), SAGD-SDA. First, the authors developed a new asphaltene precipitation model that involved four oil pseudo-components (Deasphalted Oil, Heavy Fraction, and Soluble and Solid Asphaltenes) and three pseudo-chemical reactions. Second, the asphaltene precipitation model was calibrated by a history match of a propane flood displacement experiment. Next, the cumulative oil production for a well-pair of SAGD-SDA was simulated using the asphaltene precipitation model. This simulation included a Base Case (steam-only) and two Technology Cases featuring 10 and 50% propane in cold-water equivalent steam, within a typical heavy oil reservoir (Ovalles 2017 and Ovalles 2019). **Figure 5.1** illustrates that the presence of propane accelerated oil production (**Figure 5.1A**) and increased the American Petroleum Institute (API) gravity (**Figure 5.1B**) of the crude oil compared to the steam-only case. In the scenario with 50% propane in the steam, the model indicated that the oil quality improved significantly, making it transportable by pipeline (Ovalles 2017). These results demonstrate the potential of the hybrid SAGD-SDA process for exploiting the vast reserves of heavy and extra-heavy oils available in the world (Ovalles 2017 and Ovalles 2019).

Another example is the "ESEIH". As discussed in Section 3.3.3, this technology involves pre-heating the mineral formation by using electromagnetic radiation (Radiofrequency) and injecting a solvent in a SAGD well configuration (Trautman 2014). Massarweh and Abushaikha proposed developing hybrid technologies, including biochemical-CO_2 EOR operations, smart waterflooding, green nanoparticles for EOR, and integrating renewable energy sources, such as solar, to power EOR operations (Massarweh 2024). Additionally, Sun and coworkers also suggested integrating new energies into heavy oil thermal recovery, as well as hybrid approaches such as steam injection and fire flooding technology, and Subsurface Upgrading and viscosity reduction (Sun 2024).

Demir et al. (2025) explored the feasibility of producing hydrogen directly from different types of crude oils and bitumen by combining microwave heating (MW) with *in situ* combustion. They discovered that MW-assisted fire flooding resulted in earlier gas evolution and higher hydrogen yields, especially from heavier oils and bitumen, when compared to traditional combustion methods (Demir 2025).

Several new hybrid concepts can be envisioned to increase oil recovery and decrease the environmental impacts of HCO/B production processes. Eco-friendly technologies can be implemented simultaneously or sequentially. Below is a non-exhaustive list of some of these concepts:

- Steam Injection (CSS or steamflooding) + Carbon Capture and Storage (surface or geological)
- Microwave/Radiofrequency heating + Subsurface SDA
- Cyclic steam stimulation or steamflooding + Subsurface SDA
- SAGD + Subsurface Catalytic Upgrading
- Nanoparticles + Surfactants for Chemical-Enhanced Oil Recovery (CEOR)

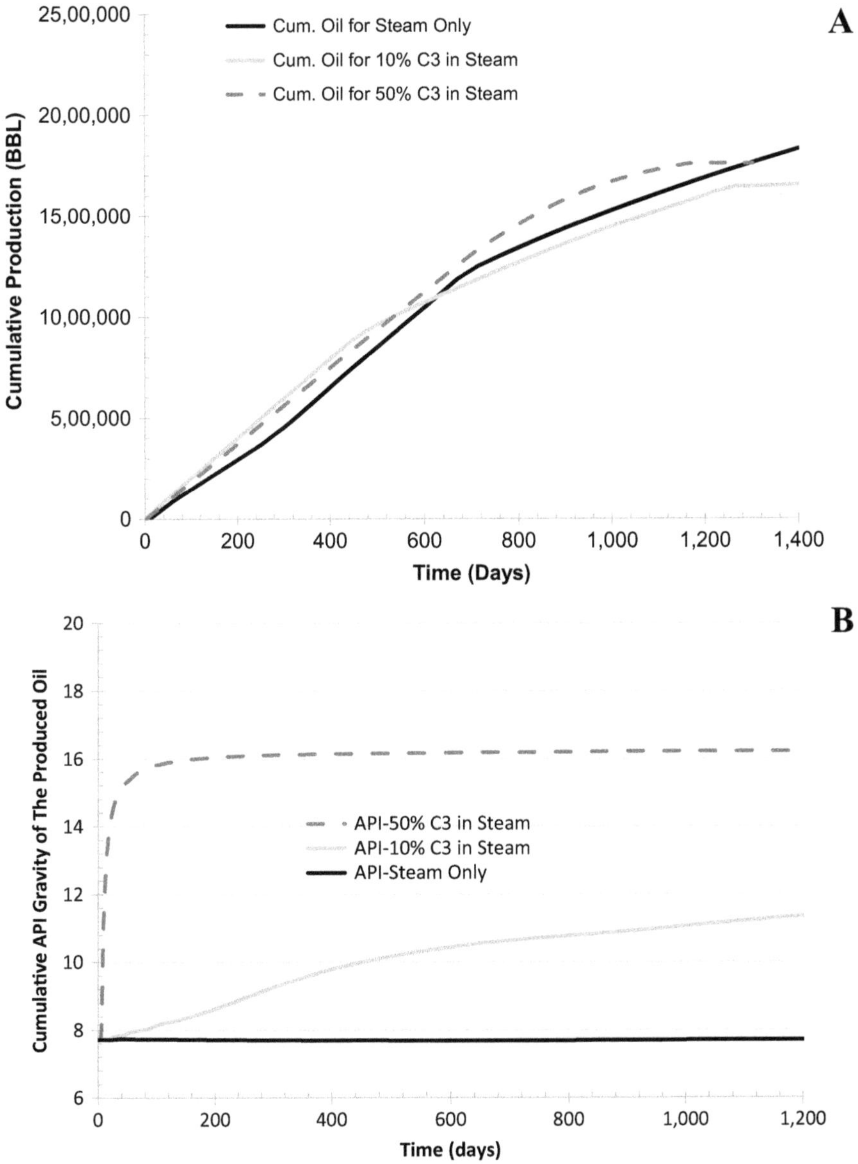

FIGURE 5.1 Numerical simulation of the hybrid concept SAGD-Solvent Deasphalting (SDA). (**A**) Cumulative oil production vs. time. (**B**) Cumulative API gravity of the produced oil.

Data taken from Ovalles (2017).

- Nanoparticles + Polymers for CEOR
- Ultrasound Applications + ASP Flooding
- Ionic Liquid injection combined with Surfactants or Nanoparticles EOR
- Toe-to-Heel Air Injection (THAI)/Catalytic Upgrading Process In Situ (CAPRI) + Carbon Capture and Storage (geological)

5.3 FUTURE DIRECTIONS

Several key strategies should be implemented to advance and develop innovative processes for the environmentally friendly production of HCO/B. First, creating advanced monitoring and modeling techniques can provide real-time insights into subsurface processes, enhancing efficiency and reducing CIs. Second, funding collaborative research can accelerate the development and improvement of new approaches toward a sustainable future. Third, it is essential to establish supportive policies, incentives, and regulations to facilitate the deployment and commercialization of these technologies. This section addresses some future research directions for reducing GHGs emissions and water use.

5.3.1 GHG Emission Reductions

In addition to hybrid concepts, numerous research efforts have explored EOR of HCO/B and bitumen and decreased the GHGs emitted into the atmosphere. Several authors have suggested that low-carbon-intensity energies, such as solar, nuclear, and geothermal, are complementary to the oil and gas industry and their use will increase in the future (Halabi 2015, Canbaz 2021, and Temizel 2022). As discussed in Section 3.4 of this monograph, these energy sources can be utilized to generate heat and electricity in oil-producing regions, treat produced water, and transport and refine crude oils. This vision underscores the investments made by major oil and gas companies to decarbonize fossil fuel production and reduce GHG emissions. In the case of solar energy, extensive R&D is still needed to develop reliable, stable thermal and storage systems that ensure a continuous supply throughout the day. The intermittent nature of solar energy makes it challenging for the traditional industry to accept (Halabi 2015).

As discussed in Section 4.3.1, microbial-derived surfactants and polymers offer a promising approach for sustainable HCO/B production. However, significant challenges must be adequately addressed, such as their vulnerability to harsh reservoir conditions (Quraishi 2021). Genetically engineered microbial EOR and enzyme EOR are emerging areas of industrial interest. In the coming decades, more R&D efforts will focus on these technologies to demonstrate their technical feasibility and improve their economic benefits. Abe and coworkers stated that research in biomaterial technology is advancing rapidly, with new functional biodegradable materials being developed and assessed for their practicality and potential to address the environmental and economic challenges facing today's world (Abe 2022). Scientific breakthroughs in this field are coming over the horizon, promising to contribute to a sustainable circular economy alongside technological advancements. Biomaterials could play a crucial role in achieving this goal (Abe 2022).

Artificial intelligence (AI) is increasingly being applied to heavy oil production, offering powerful tools to enhance efficiency, reduce costs, and lower environmental impacts (Povoas 2025). By leveraging machine learning and advanced analytics, AI could optimize reservoir modeling, predict steam-to-oil ratios, and improve the performance of environmentally friendly EOR techniques. These applications not only improve recovery rates but also help reduce CI, aligning production with sustainability goals and regulatory requirements. More R&D is expected in the future.

There is little doubt that heavy oils and bitumen will continue to play a role in the energy outlook. The key to sustainable growth and enhanced economic viability lies in developing even more efficient processes to sustain a continuous supply of environmentally responsible energy. The abundance of heavy oil and bitumen resources, combined with suitable economic policies and a stable political climate in many regions around the world, will encourage oil companies to make the substantial financial investments needed to participate in future energy development for the remainder of the century. Additionally, these innovative eco-friendly technologies could be valuable for recovering other energy resources, such as conventional oil, shale oil, and coal deposits.

5.3.2 Water Usage

Although many strategies to reduce GHG emissions from the HCO/B sector have been proposed, few evaluations have examined the impacts on water use thoroughly. Based on projected production trends, the exploitation of HCO/B will demand up to 1.31 km^2 of freshwater per year and may result in the loss of approximately 8,700 km^2 of forest (Rosa 2017). Therefore, substantial efforts should be made to mitigate the environmental impacts of HCO/B production while also meeting the energy demands necessary for sustaining global economic prosperity and quality of life.

Furthermore, a significant portion of the GHG emissions from HCO/B production is used to treat water and convert it to steam. Also, water is a crucial resource used to separate bitumen from the petroleum-sands mix, and most of the water used in this recovery process is recycled. Since 2016, the water recycling rate has increased and now accounts for 88% of the total water used for HCO/B production (Coraca 2026). However, the quality of the recycled water diminishes over time, as the recycling process continually introduces chemicals from HCOs into the steam, which ultimately reduces the oil-water recovery rate. Thus, R&D should focus on achieving water conservation and reducing GHG emissions as new, environmentally friendly HCO/B processes are developed in the future.

REFERENCES

Abe, A.A., Rossi, C.O., Caputo, P., 2022, "Biomaterials and Their Potentialities as Additives in Bitumen TAU: echnology: A Review," *Molecules*, 27, 8826–8849. https://doi.org/10.3390/molecules27248826

Agi, A., Junin, R., Jaafar, M.Z., Sidek, M.A., Yakasai, F., Gbadamosi, A., Oseh, J., 2022, "Laboratory Evaluation to Field Application of Ultrasound: A State-Of-The-Art Review on the Effect of Ultrasonication on Enhanced Oil Recovery Mechanisms," *J. Ind. Eng. Chem.*, 110, 25 June, 100–119. https://doi.org/10.1016/j.jiec.2022.03.030

Ahmadi, M., 2023, *Sustainable In-Situ Heavy Oil and Bitumen Recovery* Gulf Professional Publishing, Cambridge, MA, USA.

Alberta Government, 2025, "Alberta Oil Sands Greenhouse Gas Emission Intensity Analysis" *Open Data*, June - Open Government. Retrieved December 22, 2025.

Alsabagh, A.M., Salem, A.M., Abdel-Aleim, G.M., Khalil, S.A., Roushdi, M., 2023, "Review of Recent Developments in Enhanced Oil Recovery: The Integration of Steam Injection with Chemical Additives and their Effect on Heavy Oil Recovery", *J. Pet. Mining Eng.*, 25 (1), 19–34.

Arnold, W.A., Graham, R.E., Wagg, B.T., 2012, "Enhanced CHOPS Using SuperSump to Reduce Environmental Footprint and Increase Oil Recovery," Paper presented at the *SPE Heavy Oil Conference Canada*, Calgary, Alberta, Canada, June. https://doi.org/10.2118/157940-MS

Ashrafi, O., Zamor, O., Navarri, P., 2021, "Impact of Carbon Capture Technologies on GHG Emissions from Oil Sands In-Situ Facilities: A System Prospective," *Appl. Ther. Eng.*, 188, 116603–116616. https://doi.org/10.1016/j.applthermaleng.2021.116603

Badruzzaman, A., Hedges, J., Demayo, T., Sigworth, H., 2008, "Nuclear Energy for Unconventional Fossil-Fuel Resource Recovery," Paper presented at the *SPE Annual Technical Conference and Exhibition*, Denver, Colorado, USA, September. https://doi.org/10.2118/116294-MS

Beckman, M., 2018, "Cyclic Solvent Process Pilot", Non-Confidential Final report for emissions reduction Alberta, Project Completion Date: June 30. https://www.eralberta.ca/wp-content/uploads/2019/01/CSP-Non-Confidential-Report.pdf. Retrieved March 3, 2025.

"Boiler – Efficiency. Combustion gross and net calorific value," 2025, Boiler - Efficiency. https://www.engineeringtoolbox.com/boiler-efficiency-d_438.html. Retrieved November 18, 2025.

California Air Resources Board, 2023, "Calculation of 2023 Crude Average Carbon Intensity Values". Retrieved November 10, 2025

"Canadian oil sands achieve record output with minimal emissions increase, S&P Global reports", 2025, October 28. Retrieved November 18, 2025.

Canbaz, C.H., Temizel, C., Palabiyik, Y., Kor, K., Hendrandingrat, L., Bahar Hosgor, F., Aksahan, F., 2021, "Environmentally-Friendly Production and Recovery Processes for Heavy Presented at *SPE Trinidad and Tobago Section Energy Resources Conference* held virtually 28–30 June 2021. https://doi.org/10.2118/200897-MS

Cano, N.A., Céspedes, A., Redondo, J., Foo, G., Jaramillo, D., Martinez, D., Gutiérrez, M., Pataquiba, J., Rojas, J., Cortés, F.B., Franco, C.A., 2022, "Power from Geothermal Resources as a Co-product of the Oil and Gas Industry: A Review," *ACS Omega*, 7, 40603–40624. https://doi.org/10.1021/acsomega.2c04374

"Carbon Intensity Methodology," 2025, https://www.energyintel.com/wcod-carbon-intensity-methodology. Retrieved March 6, 2025

Cárdenas, J.C., Jaimes, M., Dorado, R., Galvis, H., Conde, O.R., Prada, A., 2024, "Application of Nanofluids for Altering Wettability in Heavy Oil: A Field Test." Paper presented at the *SPE International Conference and Exhibition on Formation Damage Control*, Lafayette, Louisiana, USA, February. https://doi.org/10.2118/217924-MS

Castellanos-Diaz, O., Verlaan, M.L., Hedden, R., 2016, "Solvent Enhanced Steam Drive: Results from the First Field Pilot in Canada", SPE No. 179815, presented at the *SPE EOR Conf. Oil& Gas west Asia*, Muscat, Oman, 21–23 March. https://doi.org/10.2118/179815-MS

Chakraborty, R., Dutta, R., Bajpai, P., Kumar, S., Kundu, G., Yomdo, S., Borgohain, T., Mandal, A., 2025, "Optimization of Control Parameters in Alkaline-Surfactant-Polymer (ASP) Flooding for Enhanced Oil Recovery: A Simulation Approach," *J. Pet. Expl. Prod. Tech.*, 15, 92–119, https://doi.org/10.1007/s13202-025-01977-x

Chen, Y., Dadashi, Z., Sleep, S., Brandt, A.R., MacLean, H.L., Bergerson, J.A., 2019, "Life Cycle Analysis (LCA) of GHG Emissions of Oil Sands Reference Facilities and Emerging Technologies" Final Report Phase 1, Submitted To: Alberta Innovates, Emissions Reduction Alberta. December 10th. https://albertainnovates.ca/wp-content/uploads/2021/01/LCA-Oil-Sands-Final-Report-December-10th-2019.pdf. Retrieved June 26, 2025.

Cheng, L., Liu, H., Huang, S., Wu, K., Chen, X., Wang, D., Xiong, H., 2018, "Environmental and Economic Benefits of Solvent-Assisted Steam-Gravity Drainage for Bitumen Through Horizontal Well: A Comprehensive Modeling Analysis," *Energy (Oxford)*, 164, 418–431. https://doi.org/10.1016/j.energy.2018.08.216

CHOPS: Cold heavy oil production with sand, 2025, https://onepetro.org/spe/general-information/1364/Cold-heavy-oil-production-with-sand. Retrieved November 10, 2025.

CHOPS: Cold heavy oil production with sand in the Canadian heavy oil industry, 2002, "Chapter 10 - Environmental Aspects and Regulations", https://open.alberta.ca/publications/2815953#summary. Retrieved October 31, 2025.

Clean Air Task Force, 2019, "CO_2 EOR Yields a 37% Reduction in CO_2 Emitted Per Barrel of Oil Produced" June 7, 2019, Clean Air Task Force. Retrieved November 4, 2025.

Clean Surfactant Energy (CSE) Project, 2023, "Cyclic Steam Replacement," December, https://www.arb.ca.gov/lists/com-attach/5106-lcfs2024-USIBYl0yAzUDd1A3.pdf. Accessed May 19, 2025

ConocoPhillips, 2017, "Catching Thieves in the Oil Sands", November 2, https://www.conocophillips.com/sustainability/sustainability-news/story/catching-thieves-in-the-oil-sands/. Retrieved June 27, 2025

Coraca, G.M., Davis, M., Kumar, A., 2026, "An Assessment of the Long-Term Water, Greenhouse Gas, and Cost Impacts Of Low-Carbon *In Situ* Oil Sands Technologies," *Fuel*, 403, 136028–136045. https://doi.org/10.1016/j.fuel.2025.136028

Demir, A.B., Albabawat, Z.W., Hajiyev, M., Karabayanova, L., Ye, H., Hascakir, B. "Hydrogen at the Source: Unlocking Clean Energy from Bitumen with Advanced Thermal Recovery." Paper presented at the *SPE Energy Transition Symposium*, Sugar Land, Texas, USA, September 2025. https://doi.org/10.2118/228353-MS

Dong, X., Liu, H., Chen, Z., 2021 "Introduction to Hybrid Enhanced Oil Recovery Processes", *Dev. Pet. Sci.*, 73, 1–46, https://doi.org/10.1016/B978-0-12-823954-4.00007-2

Dupuis, G., Niewerf, J., 2020, "A Cost-Effective EOR Technique To Reduce Carbon Intensity With Polymer Flooding and Modular Skids," *J. Pet. Tech.*, August 25th, https://jpt.spe.org/cost-effective-eor-technique-reduce-carbon-intensity-polymer-flooding-and-modular-skids

Farajzadeh, R., Glasbergen, G., Karpan, V., Mjeni, R., Boersma, D.M., Eftekhari, A.A., Casquera Garcia, A.C., Bruining, J., 2022, "Improved Oil Recovery Techniques and Their Role in Energy Efficiency and Reducing CO_2 Footprint of Oil Production," *J. Cleaner Prod.*, 369, 13308–133319, https://doi.org/10.1016/j.jclepro.2022.133308

Franco, C.A., Lina, M., Salinas, L.M., Luis, G., Alzate, L.G., Daniela Molina, D., Gabriel, J., Rendón, G.J., Cristian, C., Obregón, C.C., Sergio, H., Lopera, S.H., Oscar, E., Medina, O.E., Farid, B., Cortés, F.B., 2025, "First Field Application of Functionalized Nanoparticles-Based Nanofluids in Thermal Enhanced Oil Recovery: From Laboratory Experiments to Cyclic Steam Stimulation Process," *Fuel*, 382, Part B, 133736–133751. https://doi.org/10.1016/j.fuel.2024.133736

Goavec, M., Salaün, M., Morvan, M., Delamaide, E., Rousseau, D., Negre, A., 2023, "Towards Sustainable Oil Production: CO_2 Footprint Assessment of EOR Surfactant-Based Processes." Paper presented at the *SPE International Conference on Oilfield Chemistry*, The Woodlands, Texas, USA, June 2023. https://doi.org/10.2118/213813-MS

Gomaa, S., Salem, K.G., El-Hoshoudy, A.N., 2024, "Enhanced Heavy and Extra Heavy Oil Recovery: Current Status and New Trends," *Petroleum*, 10 (3), 399–410. https://doi.org/10.1016/j.petlm.2023.10.001

Gordon, D., Brandt, A., Bergerson, J., Koomey, J., 2015, Know Your Oil: Creating a Global Oil-Climate Index Carnegie Endowment for International Peace. Figure 3. Accessed May 18, 2025

Halabi, M.A., Al-Qattan, A., Al-Otaibi, A., 2015, "Application of Solar Energy in the Oil Industry-Current Status and Future Prospects," *Renew. Sustain. Energy Rev.*, 43, 296–314. https://doi.org/10.1016/j.rser.2014.11.030

Hama, S.M., Manshad, A.K., Ali, J.A., 2023, "Review of the Application of Natural Surfactants in Enhanced Oil Recovery: State-of-the-Art and Perspectives," *Energy Fuels*, 37, 10061–10086. https://doi.org/10.1021/acs.energyfuels.3c01347

Hamidi, H., Haddad, A.S., Otumudia, E.W., Rafati, R., Mohammadian, E., Azdarpour, A., Giles Pilcher, W.G., Fuehrmann, P.W., Sosa, L.R., Cota, N., García, D.C., Ibrahim, R.M., Damiev, M., Tanujaya, E., 2021, "Recent Applications of Ultrasonic Waves in Improved Oil Recovery: A Review of Techniques and Results," *Ultrasonics*, 110, 106288. https://doi.org/10.1016/j.ultras.2020.106288

Hannouf, M., Assefa, G., Gates, I., 2022, "Policy Insights to Accelerate Cleaner Steam-Assisted Gravity Drainage Operations," *Energies*, 15, 86. https://doi.org/10.3390/en15010086

Hoque, M., 2023 "Techno-Economic Analysis and GHG Emissions' Footprints of Solvent-steam-based Bitumen Extraction", A thesis submitted in partial fulfillment of the requirements for the degree of Master of Science, Department of Mechanical Engineering, University of Alberta

Huc, A., Argillier, J.-F., Renard, G., Quignard, A., Alazard, N., 2011, *Heavy Crude Oils; From Geology to Upgrading, an Overview*, Editions Technip, Paris.

International Energy Agency, 2024, "World Energy Out Look", Chapter 6. Retrieved December 10, 2025.

Islam, M.R., 2020, *Economically and Environmentally Sustainable Enhanced Oil Recovery*, Wiley, Hoboken, NJ, USA.

Israel, B., 2016, Measuring Oilsands Carbon Emissions Intensity," Pembina Institute, August. https://www.pembina.org/reports/measuring-carbon-emission-intensity.pdf. Retrieved October 31, 2025.

Leaute, R.P., Carey, B.S., 2007, "Liquid Addition to Steam for Enhancing Recovery (LASER) of Bitumen with CSS: Results from the First Pilot Cycle", *J. Can. Pet. Tech.*, 46 (9), 22–30. https://doi.org/10.2118/07-09-01

Lee, J., Babadagli, T., 2021, "Mitigating Greenhouse Gas Intensity Through New Generation Techniques During Heavy Oil Recovery," *J. Clean Prod.*, 286, 124980. https://doi.org/10.1016/j.jclepro.2020.124980

Maio, Y.C., Gribik, A.M., Patterson, M.W., Boardman, R.D., 2010, "Nuclear-Integrated Oil Sands Recovery via Steam-Assisted Gravity Drainage" Technical Evaluation Study, Project No. 23843. Idaho National Laboratory. Retrieved December 10, 2025.

Martinez-Palou, M. Likhanova, N.V., 2023, *Applications of Ionic Liquids in the Oil Industry: Towards a Sustainable Industry*, Bentham Sci. Pub. Pte. Ltd., Singapore, ISBN: 978-981-5079-57-9.

Masnadi, M.S., El-Houjeiri, H.M., Schunack, D., Li, Y., Englander, J.G., Badahdah, A., Monfort, J.-C., Anderson, J.E., Wallington, T.J., Bergerson, J.A., Gordon, D., Koomey, J., Przesmitzki, S., Azevedo, I.L., Bi, X.T., Duffy, J.E., Heath, G.A., Keoleian, G.A., McGlade, C., Meehan, D.N., Yeh, S., You, F., Wang, M., Brandt, A.R., 2018, "Global Carbon Intensity of Crude Oil Production," *Science*, 361(6405) 851–853. https://doi.org/10.1126/science.aar6859

Massarweh, O., Abushaikha, A.S., 2024, "Towards Environmentally Sustainable Oil Recovery: The Role of Sustainable Materials," *Energy Reports*, 12, 95–119. https://doi.org/10.1016/j.egyr.2024.06.013

MEG Energy Corp, 2024, "Enhanced Modified VAPour EXtraction R&D Operation", December 23. Retrieved June 11, 2025.

"Methodology for calculating CO_2 emissions from energy use", 2006. Retrieved November 17, 2025

Mierez, J., AlTammar, M.J., Alruwaili, K.M., Alfa, R.T., 2024, "Recent Advances of Ultrasound Applications in the Oil and Gas Industry", *Ultrasonics Sonochem.*, 103, 106767–106776. https://doi.org/10.1016/j.ultsonch.2024.106767

Moradi, M., García, W., Amado, P.M., Konopczynski, M.R., 2023, "Delivering NET ZERO–A Case Study of Minimized Carbon Intensity Production Using Autonomous Inflow Control Technologies from a Remote Location in the Peruvian Amazon."

Paper presented at the *SPE Europe- Europe Energy Conference featured at the 84th EAGE Annual Conference & Exhibition*, Vienna, Austria, June. https://doi.org/10.2118/214343-MS

Morris, P., 2021 "Effective Solvent Extraction Incorporating Electromagnetic Heating (ESEIEH™)" Project Fibal Report to the Emissions Reduction Alberta, Submission date 2022. Retrieved July 28, 2021.

Okoniewski, M., 2025, Acceleware Chief Scientific Officer & Co-Founder, personal communications, September 16th.

Oni, A.O., Monsalve, I.C.T., Kumar, A., 2023, "The Development of Life Cycle Greenhouse Gas Emission Footprints of Novel Pathways for the Solvent-Assisted and Solvent-Electromagnetic Heating Oil Sands Extraction Processes," *Fuel*, 346, 128399–128414. https://doi.org/10.1016/j.fuel.2023.128399

Orellana, A., Laurenzi, J., MacLean, H.L., Bergerson, J.A., 2018 "Statistically Enhanced Model of In Situ Oil Sands Extraction Operations: An Evaluation of Variability in Greenhouse Gas Emissions," *Env. Sci. & Tech.*, 52 (3), 947–954. https://doi.org/10.1021/acs.est.7b04498

Ovalles, C., 2019, *Subsurface Upgrading of Heavy Crude Oils and Bitumen*, CRC Press, Boca Raton: Taylor & Francis Group, https://doi.org/10.1201/9781315181028

Ovalles, C., Rogel, E., Alboudwarej, H., Inouye, A., Benson, I.P., Vaca, P., 2017, "Physical and Numerical Simulations of Subsurface Upgrading by Use of Solvent Deasphalting in a Heavy-Crude-Oil Reservoir," *SPE Res., Eval. & Eng.*, 20, 654–668. https://doi.org/10.2118/183636-PA

Pérez, R.A., Rodriguez, H.A., Rendón, G.J., Plata, B.G., Salinas, L.M., Barbosa, C., García, L.E., Rojas, F.A., Orrego, J.A., León, L.J., St. Bernard, J. G., Manrique, E. J., 2022, "Optimizing Production Performance, Energy Efficiency and Carbon Intensity with Preformed Foams in Cyclic Steam Stimulation in a Mature Heavy Oil Field: Pilot Results and Development Plans." Paper presented at the *SPE Improved Oil Recovery Conference*, Virtual, April. https://doi.org/10.2118/209399-MS

Pinilla, A., Asuaje, M., Ratkovich, N., 2025, "Comparison Between AICV, ICD, and Liner Completions in the Displacement Front and Production Efficiency in Heavy Oil Horizontal Wells" *Processes*, 13, 1576. https://doi.org/10.3390/pr13051576

Povoas, M.D.S., Moreira, J.F., Neto, S.V.M., Carvalho, C.A.D.S., Cezario, B.S., Guedes, A.L.A., Lima, G.B.A., 2025, "Artificial Intelligence in the Oil and Gas Industry: Applications, Challenges, and Future Directions," *Appl. Sci.*, 15, 7918. https://doi.org/10.3390/app15147918

Quintero, H.I., Solorzano, P.L., Barbosa, C., Corredor, L.M., Martinez, A., Feriz, E.F., Hernandez, R., Vega, S.M., Carrasca, L.K.J., Zapata, J.F., Guerrero, C.M., Dueñas, D., Rojas, C., 2024, "Carbon Footprint and Energy Intensity Assessment for Enhanced Oil Recovery (EOR) Based on Polymer Injection: A Colombian Case Study." Paper presented at the *SPE Improved Oil Recovery Conference*, Tulsa, Oklahoma, USA, April. https://doi.org/10.2118/218169-MS

Quraishi, M., Bhatia, S.K., Pandit, S., Gupta, P.K. Rangarajan, V., Lahiri, D., Varjani, S., Mehariya, S., Yang, Y.-H., 2021, "Exploiting Microbes in the Petroleum Field: Analyzing the Credibility of Microbial Enhanced Oil Recovery (MEOR)" *Energies*, 14, 4684–4714. https://doi.org/10.3390/en14154684

Rosa, L., Davis, K.F., Rulli, M.C., D'Odorico, P., 2017 "Environmental Consequences of Oil Production from Oil Sands," *Earth's Future*, 5, 158–170, https://doi.org/10.1002/2016EF000484

Safaei, M., Oni, A.O., Gemechu, E.D., Kumar, A., 2019, "Evaluation of Energy and GHG Emissions' Footprints of Bitumen Extraction Using Enhanced Solvent Extraction Incorporating Electromagnetic Heating Technology", *Energy*, 186, 115854–115866. https://doi.org/10.1016/j.energy.2019.115854

Salam, A.H., Nazar, M., Kamal, M.S., Ahad Mahboob, A., Hussain, S.M.S., Aljawad, M.S., Raza, A., Patil, S., 2025, "A Perspective On Ionic Liquids As Multifunctional Agents in Enhanced Oil Recovery and CO_2 Sequestration in Carbonate Formations," *Carbon Capture Sci. Tech.*, 15, 100043–100052. https://doi.org/10.1016/j.ccst.2025.100443

Sanchez, J., 2022, "Field Pilot of Steam Additives Technology to Reduce GHG Emissions Intensity in SAGD Final Non-Confidential Outcomes Report", December 22. Retrieved June 2, 2025

Sandler, J., Fowler, G., Cheng, K., Kovscek, A., 2012, "Solar-Generated Steam for Oil Recovery: Reservoir Simulation, Economic Analysis, and Life Cycle Assessment," Paper presented at the *SPE Western Regional Meeting*, Bakersfield, California, USA, March. https://doi.org/10.2118/153806-MS

Sawarkar, A.N., Pandit, A.B., Shriniwas, D.S., Joshi, J.B., 2009, "Use of Ultrasound in Petroleum Residue Upgradation", *Can. J. Chem. Eng.*, 87, 329–342. https://doi.org/10.1002/cjce.20169

Seidy-Esfahlan, M., Tabatabaei-Nezhad, S.A., Khodapanah, E., 2024, "Comprehensive Review of Enhanced Oil Recovery Strategies for Heavy Oil and Bitumen Reservoirs in Various Countries: Global Perspectives, Challenges, and Solutions," *Heliyon*, 10 e37826. https://doi.org/10.1016/j.heliyon.2024.e37826

Shafiai, S.H., Gohari, A., 2020 "Conventional and Electrical EOR Review: The Development Trend of Ultrasonic Application in EOR" *J. Pet. Expl. Prod. Tech.*, 10, 2923–2945. https://doi.org/10.1007/s13202-020-00929-x

Shaikhah, D., Loise, V., Angelico, R., Porto, M., Calandra, P., Abe, A.A., Testa, F., Bartucca, C., Oliviero Rossi, C., Caputo, P., 2024, "New Trends in Biosurfactants: From Renewable Origin to Green Enhanced Oil Recovery Applications," *Molecules*, 29, 301. https://doi.org/10.3390/molecules29020301

Sleep, S., Dadashi, Z., Chen, Y., Brandt, A.R., MacLean, H.L., Bergerson, J.A., 2021, "Improving Robustness of LCA Results Through Stakeholder Engagement: A Case Study of Emerging Oil Sands Technologies" *J. Cleaner Prod.* 281, 125277–125290. https://doi.org/10.1016/j.jclepro.2020.125277

Soiket, Md.I.H., Oni, A.O., Kumar, A., 2019, "The Development of a Process Simulation Model for Energy Consumption and Greenhouse Gas Emissions of a Vapor Solvent-Based Oil Sands Extraction and Recovery Process," *Energy*, 173, 799–808, https://doi.org/10.1016/j.energy.2019.02.109

Sonoro Signal, 2024 "Taking CHOPS Horizontal in Multi-Lined, Multilateral Wells (Part 1)", June 4. Accessed April 4, 2025.

"Steamflood Enhanced Oil Recovery Drives Increased Production at Duri Field", 2018, Jakarta, Indonesia, July. Retrieved 17, 2025.

Sun, H., Wang, H., Cao, X., Shu, Q., Fan, Z., Wu, G., Yang, Y., Wu, Y., 2024, "Innovations and Applications of the Thermal Recovery Techniques For Heavy Oil," *Energy Geosci.*, 5, 100332–100344. https://doi.org/10.1016/j.engeos.2024.100332

Taber, J.J., Martin, F., Seright, R., 1997, "EOR Screening Criteria Revisited-Part 1: Introduction to Screening Criteria and Enhanced Recovery Field Projects," *SPE Reservoir Engineering*, 12(03) 189–198. https://doi.org/10.2118/35385-PA

Temizel, C., Yegin, C., Hosgor, F.B., Aydin, H., Kabir, S., 2022, "Exploring Green Energy Applications in the Oil and Gas Industry, "Paper presented at the *SPE Canadian Energy Technology Conference*, Calgary, Alberta, Canada, March. https://doi.org/10.2118/208900-MS

Tileuberdi, N., Gussenov, I., 2024, "Review on Miscible, Immiscible, and Progressive Nitrogen Injection for Enhanced Oil Recovery", *Energy Rep.*, 12, 360–367. https://doi.org/10.1016/j.egyr.2024.06.004

Trautman, M., Ehresman, D., Edmunds, N., Taylor, G., Cimolai, M., 2014, "Effective solvent extraction system incorporating electromagnetic heating" US Patent No. 8,776,877.

Upreti, S.P., Lohi, A., Kapadia, R.A., El-Haj, R., 2007, "Vapor Extraction of Heavy Oil and Bitumen: A Review," *Energy & Fuels*, 21, 1562–1575. https://www.engineeringtoolbox.com/water-properties-d_1573.html

Vishnumolakala, N., Zhang, J., Ismail, N.B., 2020, "A Comprehensive Review of Enhanced Oil Recovery Projects in Canada and Recommendations for Planning Successful Future EOR Projects", SPE No. 199951. Presented at the *SPE Canada Heavy Oil Technical Conference held in Calgary*, Alberta, Canada, 29 September–2 October. https://doi.org/10.2118/199951-MS

"Water Properties: Vaporization Heat vs. Temperature - Charts and Calculator" 2025. Retrieved November 18, 2025

Wilson, K., 2016, "CCEMC Project ID: H110031 H110031" Non-Confidential Final Report, sumitted on February 13. https://eralberta.ca/wp-content/uploads/2017/05/H110031-Nsolv-BEST-Field-Pilot-Final-Report-Public.pdf. Accessed May 13, 2025

Wu, T., Zhao, Y., Zhang, Y., Li, Z., Su, J., 2023, "Matching Method Between Nanoparticle Displacement Agent Size and Pore Throat in Low Permeability Reservoir," *Front. Chem.*, 11, 1289271–1289288. https://doi.org/10.3389/fchem.2023.1289271

Zabala, R., Franco, C.A., Cortés F.N., 2016, "Application of Nanofluids for Improving Oil Mobility in Heavy Oil and Extra-Heavy Oil: A Field Test," Paper presented at the *SPE Improved Oil Recovery Conference*, Tulsa, Oklahoma, USA, April. https://doi.org/10.2118/179677-MS

Zhou, Y., Yin, D., Chen, W., Liu, B., Zhang, X., 2019, "A Comprehensive Review of Emulsion and Its Field Application for Enhanced Oil Recovery," *Ener. Sci. Eng.*, 7(4), 1046–1058. https://doi.org/10.1002/ese3.354

Glossary

AI	Artificial Intelligence
AICD	Autonomous Inflow Control Devices
API	Gravity of the sample as measured by the American Petroleum Institute. See Equation 1.1
ASP	Alkali-Surfactant-Polymer
BBBL	billion barrels (BBL × 10^9) measured in US Barrels
BBS	Branched alkylbenzene sodium sulfonate
BBL	US Barrels (42 Gal or 0.159 m^3)
BBL/D	US Barrels per day
BEST	Bitumen Extraction Solvent Technology
BS&W	Basic Sediment and Water
Btu	British Thermal Units
CAD	Canadian dollars
CAPEX	Capital costs (US$)
CAPRI	Catalytic Upgrading Process In-Situ Process
CCEMC	Alberta's Climate Change and Emission Management Corporation
CEOR	Chemical-Enhanced Oil Recovery
CHOPS	Cold Heavy Oil Production with Sand
CI	Carbon Index or Carbon Intensity, generally measured in $KgCO_2e/BBL$ F
c_p	Average water heat capacity (≈4.27 kJ/kg C) in the range 25°C–190° used in Equations 5.3 and 5.4
CSP	Cyclic Solvent Project
CSI	Cyclic Solvent Injection Process
CSS	Cyclic Steam Stimulation Process
CTI	Clean Technology Inverter
DME	Dimethyl Ether
DES	Deep Eutectic Solvent
EE	Index calculated by dividing the energy consumed by the energy produced
EJ	Exajoules (10^9 MJ)
EOR	Enhanced Oil Recovery Process
eMSAGP	Enhanced Modified Steam and Gas Push
eMVAPEX	Enhanced Modified Vapor Extraction
EDH	Electrical Downhole Heaters
EE	Energy Efficiency Index (no units)
eMSAGP	Enhanced Modified Steam and Gas Push
eMVAPEX	Enhanced Modified Vapor Extraction
EEOR	Enzyme-Enhanced Oil Recovery

ES-SAGD	Expanded Solvent-Steam-Assisted Gravity Drainage (**SAGD**)
ESEIEH	Enhanced Solvent Extraction Incorporating Electromagnetic Heating process
E_{fuel}	Fuel energy (GJ/BBL) used in Equation 5.5
g CO_2e/MJ	Grams of carbon dioxide equivalent removed for MJ of crude oil
GEMEOR	Genetically Engineered Microbial-Enhanced Oil Recovery
GHG	Greenhouse Gases
h_{vap}	Average heat of vaporization of water (2196 kJ/kg at 190°C) used in Equations 5.3 and 5.4
HBA	Hydrogen Bond Acceptor
HBD	Hydrogen Bond Donor
HCO	Heavy Crude Oils (<20°API)
HO/B	Heavy Crude Oils and Bitumen
HPAM	Hydrolyzed polyacrylamide
HTGR	High-Temperature Gas-Cooled Reactor
IEA	International Energy Agency
IFT	Interfacial Tension in mN/m (dyn/cm)
IL	Ionic Liquid
iSOR	Instantaneous Steam-to-Oil Ratio
ISC	*In-Situ* Combustion
ISR	*In-Situ* Reflex Process
ISUT®	*In-Situ* Upgrading Technology Process
JIVE	Joint Implementation of Vapor Extraction Project
kg CO_2e/BBL	Kg of Carbon Dioxide Equivalent Removed for One US Barrel of Petroleum
LAAS	Linear ionic-alkyl-aryl sulfonate
LASER	Liquid Addition to Steam for Enhancing Recovery
LCA	Life Cycle Analysis
LPG	Liquefied Petroleum Gases
LTS	Linear sodium alkyl toluene sulfonate
LVF	Liquid Volume Fraction
m	Mass (kg) of steam per BBL oil in Equation 5.2
MEOR	Microbial-Enhanced Oil Recovery
MMcf/d	Millions of cubic feet per day
MW	Microwave energy
η	Boiler efficiency used in Equation 5.5
NCG	Non-Condensable Gas (e.g., methane, nitrogen or flue gas)
NGCC	Natural Gas Combined Cycle
nm	Nanometers (10^{-9} m)
NP	Nanoparticles
NPV	Net Present Value (US$)
Nsolv	A solvent deasphalted process, see www.n-solv.com
OPEX	Operating costs (US$)
OOIP	Original Oil-in-Place, generally measured in US Barrels (BBL)
OTSG	Once Through Steam Generator

O&G	Oil and Gas
PCP	Progressive Cavity Pump
PDVSA	Petróleos de Venezuela Sociedad Anónima, Venezuelan Government Oil Company
PTRC	Petroleum Technology Research Center
PVT	Pressure-Volume-Temperature
q	Energy (kJ/kg) to raise feedwater to saturated steam (per kg) used in Equations 5.3 and 5.4
R&D	Research and Development
RF	Radiofrequency energy
SAGD	Steam-Assisted Gravity Drainage
SAGP	Steam and Gas Push
SA-SAGD	Solvent-Assisted SAGD
SCF	Standard Cubic Feet
SCGT	Simple Cycle Steam Generation
SMR	Steam Methane Reforming
SOR	Steam-to-Oil Ratio
Sp. Gr. 60F/60F	Specific gravity of the crude at 60°F vs. water at the same temperature
SSU	Subsurface Upgrading
THAI	Toe-to-Heel Air Injection Process
TRL	Technology Readiness Levels, based on the US Department of Energy in 2015
VAPEX	Vapor Extraction Process
WAG	Water Alternating Gas Injection Process

Index

Pages in **bold** refer to tables.

D

E

For Product Safety Concerns and Information please contact our EU representative GPSR@taylorandfrancis.com
Taylor & Francis Verlag GmbH, Kaufingerstraße 24, 80331 München, Germany

www.ingramcontent.com/pod-product-compliance
Ingram Content Group UK Ltd.
Pitfield, Milton Keynes, MK11 3LW, UK
UKHW020204200926
475702UK00004B/599
* 9 7 8 1 0 4 1 0 6 4 5 0 3 *